A CONVERGENCE OF LIVES

Lives of Women in Science

Pnina Abir-Am, series editor

A
CONVERGENCE
OF LIVES

Sofia Kovalevskaia:
Scientist, Writer, Revolutionary

Ann Hibner Koblitz

Rutgers University Press
New Brunswick, New Jersey

First published in hardcover by Birkhauser Boston, Inc.,
1983; second printing (revised), 1988
Reprinted in hardcover by Rutgers University Press,
1993
First published in paperback by Rutgers
University Press, 1993

Library of Congress Cataloging in Publication Data

Koblitz, Ann Hibner.
  A Convergence of Lives.

  Bibliography: p.
  Includes index.
  1. Kovalevskaia, S. V. (Sófia Vasílevna), 1850–1891.
2. Mathematicians—Soviet Union—Biography.   I. Title.
QA29.K67K6 1983   510'.92'4 [B]   83-17233
ISBN 0-8135-1962-4 (cloth)   0-8135-1963-2 (pbk.)

Manufactured in the United States of America

To the memory of my father

Michael Hibner

July 9, 1911–August 20, 1983

TABLE OF CONTENTS

SERIES FOREWORD

For the greater part of the history of science, indeed of any history, the lives and works of women have not been considered suitable subjects for biography. As a genre, biography focused on individuals who led exemplary lives of great achievement, which meant, at least in the domain of science, making discoveries, inventing theories, or conducting crucial experiments. Women's almost complete absence from the history of science in both the biographical and autobiographical genre suggests either that they played no role in such scientific achievements, or that their role has been obscured by a variety of cultural, social, political, and scientific forces.

In the last decade, under the impact of new intellectual and political currents, such as the discovery of the analytic power of the category of gender and the women's movement and its legislative implications, the lives and works of women in general, and women in science, in particular, have come to enjoy an unprecedented scholarly renaissance. This renaissance revolves around an almost archeological recovery of the diversity of women as scientific contributors in observational, theoretical, or experimental sciences, while spanning an equally great diversity in time (from the Scientific Revolution to yesterday's women Nobelists) and space (from Western Europe to South America).

Ann Hibner Koblitz's *A Convergence of Lives: Sofia Kovalevskaia: Scientist, Writer, Revolutionary*, first issued in 1983, and now reprinted with a new preface by the author, has been a pioneering work, part of the first wave of biographies establishing the centrality of women's lives and work to the practice and accomplishments of science. Together with Robert Reid's *Marie Curie* (1974), Anne Sayre's *Rosalind Franklin and DNA* (1975), Elizabeth Patterson's *Mary Somerville and the Cultivation of Science* (1983), Evelyn Fox Keller's *A Feeling for the Organism: A Biography of Barbara McClintock* (1983), and, of course, Margaret Rossiter's *Women Scientists in America* (1982), Ann Hibner Koblitz's biography of Kovalevskaia opened a new territory in reevaluating the life and work of women as major historical actors in their own right, rather than as accessories to other people's lives or as "mere" daughters, wives, or mothers of scientists.

Kovalevskaia's life and work are particularly apt topics for a biography in this series, which seeks to combine scholarship in history of science with an analytic perspective on gender in culture and society. First, Kovalevskaia's distinguished work and success in obtaining a professional position in mathematics not only set an important precedent for later generations of women, but continues to serve as a major counter-example against the sexist claims that women are less suitable for mathematics, or that mathematics is an unfeminine discipline.

Second, Kovalevskaia's own personal story, that of a woman, born and educated in Russia, who had successful careers in Germany and Sweden, sheds new light on a variety of intersecting issues of scientific, historical, and social interest. On the scientific front, her work reflects the importance of enlightened male mentors in championing the ideas of highly original women scientists. Historically, Kovalevskaia's life and career in the second half of the nineteenth century reflect the increasing internationalization of mathematics at the time, as well as the progressive (yet very slow) changes in the educational policies of universities in granting advanced degrees and positions to women. Socially, Kovalevskaia's story demonstrates that women can be productive and successful mathematicians in a variety of personal situations (i.e., single, married, widow, mother), thus counteracting long-standing social prejudices which forced many women to give up either their scholarly careers or their family lives.

Third, Kovalevskaia's activism outside science—first seeking freedom in science as a revolutionary in the "nihilist" social movement of Russian intelligentsia in the 1860s, and later as novelist and playwright addressing the liberation of women from the bondage of conventional social mores—provides an excellent model of a concern for social responsibility for both men and women scientists.

Ann Hibner Koblitz's book is unique in its depth of research in primary sources, in its broad assessment of Kovalevskaia's role in both nineteenth- and twentieth-century mathematics, and in its overall sensitivity to the intellectual, cultural, social, and political context of scientific activity that shaped Kovalevskaia's pioneering life and career.

Pnina Abir-Am
Baltimore, Maryland
September 1992

PREFACE TO THE SECOND EDITION

Since the first edition of this book appeared in 1983, there has been a resurgence of interest in the life and work of Sofia Kovalevskaia. Among mathematicians, this interest is largely due to enthusiasm about modern applications of Kovalevskaia's asymptotic method—it is not all that common in the history of mathematics for the work of a specialist so long dead to have a new period of relevance for contemporary researchers.[1]

Among historians of science, Kovalevskaia has attracted attention partly because of the publication of several biographies and biographical articles,[2] and partly because the overall inter-

---

[1]See, for example, Michael Tabor, "Modern Dynamics and Classical Analysis," *Nature*, 310 (26 July 1984); and Robert Hermann, "The Geometric Foundations of the Integrability Property of Differential Equations and Physical Systems. II," *Journal of Mathematical Physics*, 25, No. 4 (1984). In 1985, the American Mathematical Society (AMS) and the Association for Women in Mathematics (AWM) sponsored three special sessions on Kovalevskaia's mathematical legacy at the AMS meeting in Amherst, Massachusetts. Immediately afterwards, the AWM and the Bunting Institute of Radcliffe College organized a Kovalevskaia Symposium in Cambridge. Papers from the AWM and AMS events were published by the AMS in 1987 in their Contemporary Mathematics series under the title *The Legacy of Sonya Kovalevskaya* (edited by Linda Keen). Besides research articles on modern advances connected to Kovalevskaia's mathematical interests, the volume also contains Roger Cooke's "Sonya Kovalevskaya's Place in Nineteenth Century Mathematics," and my "Changing Views of Sofia Kovalevskaia."

[2]In 1984, Roger Cooke's *The Mathematics of Sonya Kovalevskaya* was published by Springer-Verlag. The book is a technical account of Kovalevskaia's researches, but it is accessible to undergraduate mathematics majors as well as professional mathematicians. In addition, in 1985 Soviet Academician P. Ia. Kochina's comprehensive 1981 *Sofia Vasil'evna Kovalevskaia* was trans-

est within the profession in the history of women's contributions to scientific fields has increased greatly in the past
decade.[3] As one of the first women in modern times to achieve
full professional status and recognition within the natural sci-

lated into English under the title *Love and Mathematics: Sofya Kovalevskaya*.
These two volumes greatly add to the amount of information on Kovalevskaia's work that is available in English. For those who read Russian, Kochina
has also published a biography of Gösta Mittag-Leffler and an edition of most
of Kovalevskaia's and Mittag-Leffler's ten-year correspondence. (Unfortunately, the vast bulk of the Kovalevskaia papers from the Swedish Institut
Mittag-Leffler archives and the [former] Soviet Academy of Sciences archives
remains unpublished. However, for excerpts referring to Kovalevskaia from
Mittag-Leffler's diaries see Lars Hörmander, "The First Woman Professor and
Her Male Colleague," in *Miscellanea mathematica* [Berlin: Springer-Verlag,
1991.])

Since the first appearance of *A Convergence of Lives* I have written a series of
articles elaborating on specific aspects of Kovalevskaia's life and professional
interactions. The above-mentioned "Changing Views" chronicles the various
myths and distortions propagated by ill-informed or occasionally ill-intentioned commentators. "A Few Words on Sofia Kovalevskaia" (AWM *Newsletter*, 13, No. 2 [1983]) and "Sofia Kovalevskaia and the Mathematical
Community" (*The Mathematical Intelligencer*, 6, No. 1 [1984]) further describe
Kovalevskaia's relations with her mathematical contemporaries and expand
upon the Prix Bordin episode; these are based heavily on the correspondence
files of the Institut Mittag-Leffler. "Career and Home Life in the 1880s" (in
Uneasy Careers and Intimate Lives: Women in Science, 1789–1979, edited by P. G.
Abir-Am and D. Outram [New Brunswick: Rutgers University Press, 1987])
analyzes the choices and compromises Kovalevskaia made in her efforts to
reconcile her personal and professional life. Finally, "Science, Women, and
the Russian Intelligentsia," *Isis*, 79 (1988), sets Kovalevskaia within the context of her age cohort: the generation of nihilist women scientists of the 1860s.

[3]See, for example, *Proceedings of the International Conference on the Role of
Women in the History of Science, Technology and Medicine in the 19th and 20th
Centuries* (Veszprém, August 15–19, 1983), 2 vols. (Budapest: MTESZ, 1983);
Abstracts of the XVIIth International Congress of History of Science (University of
California, Berkeley, 31 July–8 August 1985), 2 vols. (Berkeley: Office for His-
tory of Science and Technology, 1985); Margaret W. Rossiter, *Women Scientists
in America: Struggles and Strategies to 1940* (Baltimore: The Johns Hopkins Uni-
versity Press, 1982); *Proceedings of the Southeast Asian Seminar on Women and
Science in Developing Countries* (Hanoi, 8–10 January 1987) (Seattle: The Kova-
levskaia Fund, 1987); P. G. Abir-Am and D. Outram, eds., *Uneasy Careers and
Intimate Lives: Women in Science, 1789–1979* (New Brunswick, N.J.: Rutgers
University Press, 1987); *La Mujer en la Ciencia, la Tecnología y la Medicina/
Women in Science, Technology and Medicine* (Bilingual proceedings of the First
Central American Conference on Women in Science, Technology and Medi-
cine, Managua, Nicaragua, 24–28 August 1987) (Seattle: The Kovalevskaia
Fund, 1988); Marianne Gosztonyi Ainley, ed., *Despite the Odds: Essays on Ca-
nadian Women and Science* (Montreal: Vehicule Press, 1989); G. Kass-Simon,
Patricia Farnes, and Deborah Nash, eds., *Women of Science: Righting the Record*

ence community, Kovalevskaia comes to mind whenever historians of women in science look for cross-cultural points of comparison.

But by far the greatest interest in Sofia Kovalevskaia arises from the discussion, beginning in the mid 1970s, about women's place in the natural sciences, and about the role that gender plays or might play in scientific research.[4] Increasing documentation of women's previously invisible and unsung participation in the development of the natural sciences has been paralleled, paradoxically, by an increase in theorizing about the posited incompatibility of women and the scientific enterprise.[5]

Some writers trace a historical connection of science, objectivity and masculinity back to the roots of modern science in the sixteenth and seventeenth centuries, and argue that:

> In sympathy with an increasing polarization between masculine and feminine, public and private, work and home, the rhetoric of modern science pitted mind against nature, reason against

(Bloomington: Indiana University Press, 1990); Veronica Stolte-Heiskanen et al., eds., *Women in Science: Token Women or Gender Equality?* (Oxford: Berg Publishers Limited, 1991).

[4]See, for example, Evelyn Fox Keller, *Reflections on Gender and Science* (New Haven: Yale University Press, 1985); Ruth Hubbard, Mary Sue Henifin, and Barbara Fried, eds., *Women Look at Biology Looking at Women* (Cambridge, Mass.: Schenkman Publishing Company, 1979); Carolyn Merchant, *The Death of Nature* (San Francisco: Harper & Row, 1980); Elizabeth Fee, "A Feminist Critique of Scientific Objectivity," *Science for the People*, 14, No. 4 (July/August 1982), pp. 5–8, 30–33; Ruth Bleier, *Science and Gender: A Critique of Biology and Its Theories on Women* (New York: Pergamon Press, 1984); Bonnie B. Spanier, "Women's Studies and the Natural Sciences: A Decade of Change," *Frontiers*, 8, No. 3 (1986), pp. 66–72; Ruth Wallsgrove, "The Masculine Face of Science," in *Alice Through the Microscope*, ed. Lynda Birke et al. (London: Virago Ltd., 1980), pp. 228–40; Lynda Birke, *Women, Feminism and Biology: The Feminist Challenge* (New York: Methuen, 1986); Sandra Harding, *The Science Question in Feminism* (Ithaca, N.Y.: Cornell University Press, 1986); Nancy Tuana, ed., *Feminism and Science* (Bloomington: Indiana University Press, 1989).

[5]See, for example, Leanna Standish, "Women, Work, and the Scientific Enterprise," *Science for the People*, 14, No. 5 (September/October 1982), pp. 12–18; Barbara Dodds Stanford, "Women and Science," *Science for the People*, 18, No. 1 (January/February 1986), pp. 5–9, 27; Tuana, *Feminism and Science*; Susan Bordo and Alison Jaggar, eds., *Gender/Body/Knowledge: Feminist Reconstructions of Being and Knowing* (New Brunswick, N.J.: Rutgers University Press, 1989); Mary Jacobus, Evelyn Fox Keller, and Sally Shuttleworth, eds., *Body/Politics: Women and the Discourses of Science* (New York: Routledge, 1990).

feeling, objective against subjective. In parallel with society's growing denial of female sexuality, it offered a deanimated, desanctified, and increasingly mechanized conception of nature. In so doing, science itself was an active agent of change. If an image of male prowess supported a particular vision of science, the same image also supported a new definition of manhood, one purified of all traces of the 'feminine.' The making of modern science went hand in hand with the making of modern men and women, supported and sustained by new familial and social norms.[6]

Analyses of the role that the scientific establishment has played in legitimizing gender, class, and racial inequalities are of course not unique to the present generation of feminist scholars. Ruth Bleier, Marian Lowe, Donna Haraway, and others are joining the ranks of those who have made important contributions to the evaluation of the scientific enterprise as a social phenomenon.[7] Their critiques of the lack of true objectivity in many conventional "scientific" theories in genetics, endocrinology, psychoanalysis, anthropology, embryology, and other fields associated with human and animal behavior can be very persuasive.[8]

[6]Evelyn Fox Keller, "Contending with a Masculine Bias in the Ideals and Values of Science," *The Chronicle of Higher Education*, October 2, 1985, p. 96.

[7]See, for example, Darrell Huff, *How to Lie with Statistics* (New York: Norton Books, 1954); Serge Lang, *The File* (New York: Springer-Verlag, 1981) and "The Professors: A Survey of a Survey," *The New York Review of Books*, 18 May 1978; Neal Koblitz, "A Tale of Three Equations; or The Emperors Have No Clothes," *The Mathematical Intelligencer*, 10, No. 1 (January 1988) and "Mathematics as Propaganda," in *Mathematics Tomorrow*, ed. Lynn Steen (New York: Springer-Verlag, 1981); Mark Kac, "Marginalia: Florence Nightingale among the Statisticians," *American Scientist* (January–February 1984); Stephen Jay Gould, *The Mismeasure of Man* (New York: Norton Books, 1981); Leon Kamin, *The Science and Politics of IQ* (New York: John Wiley, 1974).

[8]See, for example, Bleier, *Science and Gender*; Marian Lowe, "Sociobiology and Sex Differences," *Signs*, 4, No. 1 (August 1978), pp. 118–25; Donna Haraway, "Animal Sociology and a Natural Economy of the Body Politic, Part II," *Signs*, 4, No. 1 (August 1978), pp. 37–60; Sarah Blaffer Hrdy, *The Woman That Never Evolved* (Cambridge: Harvard University Press, 1981); Anne Fausto-Sterling, *Myths of Gender: Biological Theories about Women and Men* (New York: Basic Books, 1985). While feminist critiques of the behavioral and biological sciences are numerous and well-argued, I have yet to encounter a gender-based analysis of the content (as opposed to the style) of the physical sciences that I have found convincing.

However, some feminist theorists go much further: they maintain that the natural sciences as presently constructed are inherently masculine, and that somehow a feminist scientist is a contradiction in terms.[9] These claims are more problematic, and are supported by little evidence. Most historians who have examined concrete cases of women's participation in the sciences in various periods are hesitant to indulge in such generalizations. In her Foreword to *Uneasy Careers and Intimate Lives: Women in Science 1789–1979*, Margaret W. Rossiter points out that the twelve essays contained in the volume "demonstrate that the interactions of 'science and gender,' as the topic is commonly phrased today, have been so complex that they defy reduction to easy theories." And in their Introduction to the same work, editors Pnina G. Abir-Am and Dorinda Outram remark: "Although the philosophical perspective on scientific knowledge as gender-biased has provided an important contribution to an understanding of both scientific and feminist ideologies as constraining and enabling women's participation in science, it has been impaired by lack of empirical data on relationships between women and science in precisely delineated historical and sociopolitical contexts. It has failed to account for historical [and, I would add, cross-cultural] diversity in women's scientific experience."[10]

The questions naturally arise: How does the story of Sofia Kovalevskaia fit into the ongoing debate on the interactions of gender and science? What can her experiences tell us about the nature of science and the scientific community? Did Kovalevskaia experience philosophical or psychological conflicts doing mathematics? Was she somehow untrue to her feminine essence by becoming part of the scientific research establishment of her time?

Since the beginning of my research on Kovalevskaia, I have been fascinated by the ways in which her story has been used to prove various points—often, paradoxically, the supposed inherent unsuitability of women for scientific work. Kovalevskaia's achievements were systematically distorted in the years

[9]See, for example, Keller, *Reflections on Gender and Science*, pp. 174–75; Standish, "Women, Work, and the Scientific Enterprise," pp. 14–17.

[10]*Uneasy Careers and Intimate Lives*, pp. xii, 3.

after her death in ways that diminished her perceived importance in the history of science and obscured her value as a role model for young women seeking to enter mathematics and the sciences.[11] In fact, Kovalevskaia was so successfully disparaged that two of the greatest women scientists of the next generation—Emmy Noether and Maria Sklodowska Curie—seem to have been either unaware of her existence, or (more likely) reluctant to acknowledge her as a predecessor because of the rumors surrounding her name.[12] This kind of fictionalization and active belittlement of the accomplishments of women scientists have far-reaching effects. As Joanna Russ has pointed out (with reference to women writers), "When the memory of one's predecessors is buried, the assumption persists that there were none and each generation of women believes itself to be faced with the burden of doing everything for the first time."[13]

In the case of Sofia Kovalevskaia, we are faced with a slightly different situation. Here, the problem is not that the memory of her had faded away. Rather, instead of sinking into genteel obscurity like most nineteenth-century male mathematicians of her stature, Kovalevskaia's name and accomplishments became buried in myth and even, to some extent, in infamy. Hermann Hettner, Paul DuBois-Reymond, Carl Runge, Leo Königsberger, and Hermann Schwarz were contemporaries of Kovalevskaia, of roughly comparable mathematical accomplishments. No one claims, however, that they did not prove the theorems with which they were credited, or that they were mascots rather than participating members of the mathematical community, or that they were unhappy doing mathematics. Nor does anyone insinuate that they succeeded because they slept with their mentors. These stories have been reserved for Kovalevskaia.

[11]I chronicle this systematic disparagement in "Changing Views of Sofia Kovalevskaia," in *The Legacy of Sonya Kovalevskaya*, ed. Linda Keen (Providence, R.I.: American Mathematical Society, 1987).

[12]I would like to thank Professor Margaret Rossiter and her class on women in science at Harvard University. Their questions about Kovalevskaia's apparent lack of influence on Curie and Noether helped to clarify my thinking on this point.

[13]Joanna Russ, *How to Suppress Women's Writing* (Austin: University of Texas Press, 1983), p. 93.

Interestingly enough, a fair number of the stories have been spread by female commentators claiming to be feminists. Anne Charlotte Leffler (or Anna Carlotta, as she called herself in later life), sister of Kovalevskaia's Stockholm colleague Gösta Mittag-Leffler, is a case in point. She was known in Scandinavia as a champion of women's rights, yet her biography of Kovalevskaia was romanticized, distorted, and essentially antifeminist in tone and content.[14]

Leffler had many imitators. A typical example was the American Isabel F. Hapgood, who wrote at the end of the nineteenth century. She took Leffler's dismal portrait, exaggerated it even more, and claimed that Kovalevskaia's life "proved" that a woman's "feminine heart" would always prevent her from competing successfully with men in intellectual life. Hapgood pointed out that Kovalevskaia had once said she lectured better when Mittag-Leffler was in the audience, and cited this as proof that she "was always mentally dependent upon a man." Hapgood also declared that Kovalevskaia was incapable of original scientific thought: "Notwithstanding her solid contributions to applied mathematics, she originated nothing; she merely developed the ideas of her teachers."[15]

A Kovalevskaia legend emerged, fueled by the distortions of Anne Charlotte Leffler, Isabel Hapgood, and their imitators, and nourished in the antifeminist climate of the 1890s. Like all fabrications, the Kovalevskaia myth had several variants, and of course was inaccurate in minor details of fact as well. But essentially, the story followed the lines of this version:

> Sonia had the brains of a man; she enjoyed a European fame; she took an academic prize for mathematics in Paris. When she was old and faded, she fell in love with a French professor. He did not return the passion, and she broke her heart. Did you ever hear of a male mathematician who died at forty-five because a woman would not have him? Of what avail was Sonia's intellect when her affections overbalanced it, when she drooped

[14]*Sonya Kovalevsky, Her Recollections of Childhood with a biography by Anna Carlotta Leffler, Duchess of Cajanello* (New York: The Century Company, 1895). See also the bibliographical note at the end of this volume, p. 278.

[15]Isabel F. Hapgood, "Notable Women: Sonya Kovalevsky," *Century Magazine*, 50 (1895), p. 539.

for want of love, when she went to her grave a martyr to the destiny of her sex? In the face of that story, why pretend that the feminine temperament is a myth, and that women are every whit as strong as men? . . . [the ironic spirit] drops a tear on poor Sonia's memory; and it reflects that love and mathematics make a dangerous mixture for a woman.[16]

The nineteenth-century commentaries could perhaps be considered no more than historical oddities— curious in their way but irrelevant to current debates on questions of gender and science—were it not for their disturbing similarity to statements made by present-day writers. Compare the following quotations of then and now:

There is only one point which I should like to emphasize in these six types of modern womanhood, and that is, the manifestation of their womanly feelings. I want to show how it asserts itself in spite of everything—in spite of the theories on which they built up their lives, in spite of the opinions of which they were the teachers, and in spite of the success which crowned their efforts. . . . They were out of harmony with themselves, suffering from a conflict which made its first appearance in the world when the 'woman question' came to the fore, causing an unnatural breach between the needs of the intellect and the requirements of their womanly nature. . . . A woman who seeks freedom by means of the modern method is generally one who desires to escape from a woman's sufferings . . . but in so doing she unconsciously deprives herself of her womanliness.
—Laura Marholm Hansson, 1896[17]

What is the conclusion of the whole matter? Setting aside all partisan questions, it would seem to be this: that a masculine head united to a feminine heart is likely to prove a very unhappy combination for a woman.
—Isabel F. Hapgood, 1895[18]

[16]L. F. Austin, "At Random," *The Sketch* (London), 24 March 1897, p. 390.

[17]Laura Marholm Hansson (Mohr), *Six Modern Women: Psychological Sketches*, trans. Hermione Ramsden (Boston: Roberts Brothers, 1896), pp. v–vi. Kovalevskaia was one of the six women profiled.

[18]Isabel F. Hapgood, "Notable Women: Sonya Kovalevsky," p. 539.

Perhaps girls, like boys, sometimes seek escape from engulfing intimacy in infancy and childhood. Their struggle for isolation and mastery over people and things may lead them to deny their essential connection to others. They may find the social environment of patriarchal institutions [Standish is referring to science] a place to reaffirm their autonomy and escape the discomfort of intimate relationships.

—Leanna Standish, 1982[19]

In a science constructed around the naming of object (nature) as female and the parallel naming of subject (mind) as male, any scientist who happens to be a woman is confronted with an a priori contradiction in terms. This poses a critical problem of identity: any scientist who is not a man walks a path bounded on one side by inauthenticity and on the other by subversion. . . . Only if she undergoes a radical disidentification from self can she share masculine pleasure in mastering a nature cast in the image of woman as passive, inert, and blind.

—Evelyn Fox Keller, 1985[20]

Hansson and Hapgood might well be considered the ideological ancestors of present-day gender and science theorists. They clearly feel that the essence of femininity is somehow opposed to the essence of intellect, and that exercise of intellect results in tension and unhappiness for women. Standish and Keller would perhaps be less willing than Hapgood and Hansson to generalize about all intellectual endeavor, but they certainly would concur that for women, science and mathematics are alienating in the extreme. For all four writers, only women who can undergo, in Keller's words, "radical disidentification from self" can take pleasure in scientific work. Only women who wish to, as Standish puts it, "escape the discomfort of intimate relationships" will be happy in the world of scientific research.

Set against the diversity of experience of women in science, the above categorizations have little explanatory force. It is unfortunate that none of these commentators bothered to undertake a sympathetic study of the concrete situations of women

[19]Standish, "Women, Work, and the Scientific Enterprise," p. 16.

[20]Keller, *Reflections on Gender and Science*, pp. 174–75.

scientists of different cultures and historical periods before rushing into stereotype and generalization.

In the specific case of Sofia Kovalevskaia, the "radical feminist" analysis accords ill with the actual circumstances. Kovalevskaia's whole life, and that of her generation of nihilist women scientists, was a constant struggle against the constraints of patriarchal institutions and mores. Yet science for her was liberating, an escape from the patriarchal society she found so confining. For Kovalevskaia, being a woman and a mathematician was anything but alienating. Mathematics was a world of elegance and beauty and order, and mathematicians were her companions, allies, and supportive friends in that world.

The experiences of the first generation of Russian women to enter the sciences in the 1860s and 1870s illustrate some of the pitfalls of subscribing to a simplistic view of the interactions of gender and science. We have here what at first looks like a totally anomalous situation, certainly one that seems to have no obvious parallel in Western Europe or North America. Kovalevskaia and the other women of her circle, far from being repelled by the sciences, were attracted to them in preference to what are usually considered the more "feminine" humanities.[21] Rather than being discouraged by male scientists, the women were respected, encouraged, and given recognition in their professions. And for the most part, the women seem to have found scientific research deeply satisfying, both emotionally and intellectually.

This is not to say that the women did not encounter obstacles and discrimination; they certainly did. But they were far more likely to meet with prejudice and resistance on the part of administrators and academics in nonscientific disciplines and the public at large than they were from their scientific colleagues. And the situation of women attempting to break into nonscientific professions (law, accounting, administration, politics, finance, the military, and so on) appears to have been much worse.

[21]A fascinating example of a similar attitude toward the natural sciences can be found in Feride Acar, "Women in Academic Science Careers in Turkey," in *Women in Science*, ed. Stolte-Heiskanen, pp. 147–71.

A series of myths and stereotypes are challenged by the experiences of Kovalevskaia and the other Russian women scientists who were educated in the 1860s and 1870s. Obviously, the idea that women "naturally" prefer and have more talent for the humanities than for the sciences is brought into question. Second, in Russia, and apparently in continental Europe as well, women seem to have been able to create a place for themselves in the sciences sooner than they did in the humanities. Third, the stereotype of male scientists put forward by many feminist theorists today—that they create an ambiance that is particularly hostile to women—clearly was false during the "golden age" of Russian science.

The story of Kovalevskaia would appear to indicate that one should proceed with caution before making generalizations about the interactions of gender and science over time and across cultures. As Ruth Bloch has pointed out: "The modern history of sex roles, far from exhibiting one continuous line of development, as many scholars suggest, has proceeded in different, even contradictory directions."[22] The warnings contained in Natalie Davis's classic article on women's history seem appropriate here: "I think we would do better to use these [sexual] polarities only when our historical evidence supports them, and not assume that they always represent the fundamental meanings that society sees in the sexes. . . . It is essential that we distinguish speculation from generalization, that we know when we are working from inadequate evidence."[23]

Finally, I would like to point out a parallel that might not seem obvious at first, but which has become increasingly clear as I have studied women in science cross-culturally. Namely, since Russia in the 1860s and 1870s was a so-called "developing" country, there are comparisons that can be drawn between the situation then and that which exists in developing countries today. In many such countries, one finds that a large section of the intelligentsia sees itself as morally and socially

[22]Ruth H. Bloch, "Untangling the Roots of Modern Sex Roles: A Survey of Four Centuries of Change," *Signs*, 4, No. 2 (Winter 1978), p. 237.

[23]Natalie Zemon Davis, "'Women's History' in Transition: The European Case," *Feminist Studies*, 3, No. 3/4 (Spring/Summer 1976), pp. 92–93.

bound to liberalize, modernize, and champion the rights of dis-
enfranchised segments of society. In such circumstances the
natural sciences are seen as allies of progressive movements, as
active forces against backwardness, superstition, repression,
and bureaucracy. Progressive scientists, in turn, see women as
their natural allies, and actively support their entry into higher
education and the professions (as well as into movements for
societal reform).[24]

For Sofia Kovalevskaia and the other Russian scientists (men
as well as women) who were educated in the 1860s and 1870s,
science was creative, fruitful, beautiful, nurturing, diverse,
welcoming of innovation, and intrinsically progressive and
egalitarian. Their image of science was a far cry from the model
of modern science as hierarchical, dominating, cold, and exclusively masculine that is put forward by some writers today.
Both the stories of Russian women scientists of the nineteenth
century and those of women scientists in many developing
countries in the twentieth century support the view that more
concrete, nuanced, and cross-cultural studies of the interactions of gender and science are needed.

Sofia Kovalevskaia's life and accomplishments are of continuing relevance for women in various parts of the world today. A century after her death, her memory is still cherished,
perhaps now more than ever. In Russia, a Kovalevskaia Memorial Museum is being established on the grounds of the restored Palibino estate; the director, Valentina Rumiantseva, has
asked me to send my works and collect modern American ma-
terials mentioning Kovalevskaia for the exhibits and archives of
the museum.

In the United States, the Association for Women in Mathematics (AWM) and the Association for Women in Science
(AWIS) sponsor regional Kovalevskaia High School Days. During these programs, young women are introduced to the sto-

[24]Present-day examples of this (in Honduras and the state of Kerala in In-
dia) can be found in my "Women in Science, Technology, and Medicine—An
Overview of Issues and Research Topics," in *La Mujer en la Ciencia, la Tecnología y la Medicina*, pp. 75–77. See also other articles in that volume; *Proceedings of the Southeast Asian Seminar*; Acar, "Women in Academic Science Careers in Turkey."

ries of Kovalevskaia and other prominent women scientists, and are encouraged to think about careers in mathematics and the sciences. Such events are particularly important because of the danger that young women will be discouraged from entering the sciences as a result of the current "postfeminist" ideological climate in the U.S. (Unfortunately, the antifeminist backlash can be reinforced by the writings of certain "radical feminist" theorists on the supposed incompatibility of women and science.)[25]

In Mexico, a new association of women mathematicians and scientists, centered around the National Autonomous University of Mexico (UNAM), was recently formed. One of their first large activities was a Kovalevskaia memorial celebration, commemorating the centenary of her death in 1991. Events included historical talks on Kovalevskaia and other women scientists, reports on the status of women professionals in Mexico, and festivals aimed at attracting young women into scientific careers.

In Germany, the Universität Kaiserslautern has instituted a visiting women's professorship in applied mathematics, named in honor of Sofia Kovalevskaia. At the inauguration ceremonies in February 1992, speakers made clear both the contemporary significance of Kovalevskaia's mathematical work and the enduring value of her example as an inspiration for young women wanting to pursue scientific careers.

Finally, I should mention that the royalties from *A Convergence of Lives* go to the Kovalevskaia Fund for Women in Science in Developing Countries, a nonprofit foundation of which I am director. The Fund has sponsored two international conferences, one in Southeast Asia and one in Central America, on the role of women in science, technology, and medicine in developing countries, and has published both conference pro-

---

[25]See, for example, Camilla Benbow and Julian Stanley, "Sex Differences in Mathematical Ability: Fact or Artifact?", *Science*, 210 (1980), pp. 1262–64. Alice Schafer and Mary Gray refute this study in their editorial "Sex and Mathematics," *Science*, 211 (1981), p. 229. See also Mary Beth Ruskai, "Letter on Feminism and Women in Science," Association for Women in Mathematics *Newsletter*, 16, No. 3 (May/June 1986), pp. 4–6; and my "A Historian Looks at Gender and Science," *International Journal of Science Education*, 9, No. 3 (1987), pp. 399–407.

ceedings. In addition, the Fund is involved in other small projects to encourage the participation of women in the sciences in Asia, Africa, and Latin America. This seems a fitting way to honor the memory of Kovalevskaia, who once said:

> . . . is it really possible not to stretch out one's hand, to refuse to help someone who is seeking knowledge and cannot help herself reach its source? After all, on woman's road, when a woman wants to take a path other than the well-trodden one leading to [traditional] marriage, so many difficulties pile up. I myself encountered many of these. Therefore I consider it my duty to destroy whatever obstacles I can in the paths of others.

Oneonta, New York
12 September 1992

PREFACE

The second half of the nineteenth century was an exciting time in European intellectual and social history. The period saw the growth of revolutionary activism, the rise of Darwinian evolutionary biology, the emergence of women's rights movements, and other challenges to established ways of thinking. Progress and change were the key words of the day, and most members of the educated classes felt confident that the future would be bright.

In Russia especially, the "intelligentsia" (an amorphous, peculiarly Russian class of professors, writers, and thinkers) had the feeling that they were on the threshold of a great new age. Russia's ignominious defeat in the Crimean War in 1856 signaled to many that wide-ranging political and social reforms were urgently needed. Most of the intelligentsia hoped that the defeat would be followed by the emancipation of the serfs, modernization of education, moves toward the equality of women, and other reforms.

It was in this period that the Russian intelligentsia developed into an influential social group that concerned itself with much more than just cogitation and empty philosophizing. During the second half of the nineteenth century, Russian intellectuals increasingly assumed burdens that their counterparts in other European countries considered outside their province. The intelligentsia became the source of most political and social activism in Russia, the social conscience and often the sole voice of protest against autocratic and reactionary policies.

Russian intellectuals were in many ways more socially progressive than their Western European colleagues. They considered it their duty to chronicle the problems of Russia and direct the way toward a more just society in the future. Their attitudes toward their less fortunate fellows—serfs, peasants, and workers—were for the most part egalitarian and non-elitist. As a class they were far more concerned with women's rights than were most European intellectuals, and disputed entrenched, patriarchal views of the family and the social order.

These progressive, anti-traditional attitudes came to be seen by other Europeans as the hallmarks of the Russian intelligentsia. Indeed, if truth be told, Russian intellectuals inspired a certain amount of alarm and antagonism among many of their European counterparts. The latter had more conservative views, and felt that the task of an educated elite was to preserve class distinctions and the sexual status quo, not attempt to eradicate them.

European intellectuals and European educated society in general especially looked askance at one segment of the Russian intelligentsia which came into prominence in the late 1850s and 1860s. This group was composed of progressive thinkers, publicists, and scientists who went even further than their fellows. Taking their cue from Turgenev's hero Bazarov in *Fathers and Sons* (the usual mistranslation of *Fathers and Children*), they called themselves "nihilists."

Turgenev seems to have meant to offend the young intellectuals of the 1860s by calling them nihilists. He wanted to show that they rejected everything and respected nothing in tsarist society; that they were mindless iconoclasts. But many of the young people, far from being insulted by the term, embraced it for their own. They *were* nihilists, they declared. Nothing was sacred to their way of thinking; society needed to be rebuilt completely.

Of course, it was an oversimplification for the nihilists to claim that they denied everything. And certainly Turgenev's fear that they respected nothing was ill-founded. In fact, they had almost boundless faith in the power of education to win out against superstition and backwardness; they were confident that woman's potential was fully equal to that of man; and they had a naive belief that the natural sciences would, given free reign, conquer all of humanity's ills.

The early nihilists believed in "small deeds," like becoming a country doctor or starting a school for peasants or tutoring young women so that they could enter foreign universities. Most of all, they believed that one's first task should be self-education, preferably in the natural sciences. Only through self-development could one attain a sufficient level of moral strength and technical expertise. Only after these qualities had been acquired could one strive toward the main goal—education and material aid for the masses of ordinary Russians.

The nihilism of the Russian intelligentsia in the 1860s had little in common with the more drastic philosophical position that later bore the name. Russian nihilism in its first incarnation was non-violent. In fact, the beliefs of the movement were not incompatible with the mild form of liberalism prevalent in American academic circles today. But nevertheless the intellectuals ran into trouble because the repressive tsarist government did not tolerate even moderate forms of dissent. Many of the nihilists were exiled to Siberia for "crimes" like possessing a censored book, or signing a petition, or participating in a peaceful demonstration.

The early nihilists, or "children of the sixties," as they often called themselves, looked forward to the (peaceful) social revolution they considered inevitable. They felt that the best way to help the revolution along was through intensive study of the natural sciences. For them, the sciences were the shining example of the best in human knowledge. Science triumphed over religion and superstition, improved people's lives through its discoveries, and "proved" through the theory of evolution that progress was inevitable. For the nihilists, science was virtually synonymous with progress and truth.

Given the nihilists' faith in the natural sciences, and their frequent use of metaphors and analogies taken from the sciences to illustrate their philosophy, it is not surprising that many "children of the sixties" were attracted to the prospect of a scientific, medical, or technical career. The more people entered the sciences, they thought, the sooner would come the day of social revolution and equality for all.

As a result of their perception of the social role of the sciences, nihilists were predominantly found in the areas of chemistry, biology, mathematics, physics, medicine, geology and other scientific fields. And, largely as a result of the nihilists' enthusiasm and dedication, Russian science at this time flourished and began to gain for itself an enviable reputation in Europe.

Prior to the end of the Crimean War, science in Russia was not advanced. True, St. Petersburg was the home of an academy of sciences that was known and respected throughout Europe. It had supported such famous scientists as Euler, the Bernoullis, von Baer, and others. But the Academy was an isolated institution, drawing its membership almost exclusively from the ranks of non-Russian (mostly German) scientists. These foreign Academy members typically had contempt for their Russian colleagues, and felt their scientific attainments to be vastly superior to those of native Russian scientists and university science teachers. For the most part, they remained aloof from Russian society, taking little part in science education and the training of native specialists.

Even if they had wished to improve scientific education in Russian high schools and universities, however, the academicians would have met with opposition. Before the Crimean War, scientific and technical subjects were not priorities in the Russian school system. Science was thought to represent a challenge to the traditional, religious, authoritarian attitudes that the tsarist government sought to instill in its subjects. A curriculum which emphasized classical languages, religion, rhetoric and the like was therefore thought more suitable.

After the humiliating defeat of Russian forces in the war, however, the views of the government changed. The military argued that it had been drastically impeded by lack of men with technical expertise in engineering and ballistics, and the medical corps complained of having to make do with inadequately trained physicians and a small number of poorly prepared support personnel. Education had to be modernized, the military maintained; scientific subjects had to be added to the high school curriculum, and promising young people should be sent abroad for advanced technical education.

The tsarist government was understandably reluctant to accede to the demands. It felt that the connections of the sciences to a materialist world view and disrespect for established authority and tradition were too strong to be ignored. But the lessons of the Crimean War and the representations of military and medical experts convinced the government to liberalize and modernize the Russian educational system.

For the nihilists, the reforms were timely. Young men in large numbers went abroad to study the sciences. Many of them either were sympathetic to nihilism to start with, or were soon converted to the philosophy by their more advanced Russian comrades. Then, when these young men returned to Russia, they incorporated their materialist, progressive, egalitarian world view into their lectures on physiology, embryology, anatomy, geology, and so on.

This was an exciting time in the history of Russian science. The nihilists' enthusiasm for and faith in science, coupled with the government's decision to devote material resources to scientific development, produced rapid results: the 1860s and 1870s were a "golden age" of Russian science.

The historian of science or scientific specialist is well-acquainted with the names of Sechenov, Timiriazev, and Mechnikov in physiology; Mendeleev, Butlerov, and Markovnikov in chemistry; Aleksander Kovalevskii in embryology; and Vladimir Kovalevskii in paleontology. What is perhaps less well known is that to varying degrees all of these eminent scientists considered themselves nihilists, and imparted their philosophies to their students. (They were so successful at their proselytizing, in fact, that Ilya Mechnikov later complained that the most capable students were no longer satisfied with scientific careers—they wanted to become full-time political activists.)

This connection of nihilism and the natural sciences had far-reaching consequences. Scientists were at the forefront of movements for popular education and social reform. They spoke out against injustice (as much as they could under the repressive tsarist system) and taught their students to expect and work for a better society. They smuggled forbidden political and scientific works into Russia, and mulled over various plans to hasten along the day of equality and social justice for all.

Although these "children of the sixties" were not committed full-time revolutionary activists as some of their students would later become, nevertheless they set the stage for subsequent revolutionary movements. Moreover, throughout their lives they would show a continuing interest in radical politics, and would often give practical assistance to members of the revolutionary underground.

The intelligentsia as a whole and the early nihilists in particular performed another important function in Russia in the 1860s and 1870s: they aided in the integration of women into the intellectual and clandestine political life of the country. As was mentioned before, most of the intelligentsia believed that women were equal to men in capabilities and potential. Intellectuals thought that women had a right and a duty to educate themselves so they could help the masses of ordinary Russians. The nihilists went further. They believed that men had a moral obligation to aid the women in all possible ways.

The nihilists' expressed belief in the equality of women was not an empty platitude. For the most part, they put their ideas into practice, and women were found in surprisingly large numbers in the membership and even the leadership of nihilist and later populist revolutionary groups. Moreover, the nihilists made real attempts to change the social structure of marriage and family relations. Some of them went so far as to say that until inequalities in the social system had been redressed, women should be given priority and deferred to in matters of education, career, and family decisions.

It was during this period that women first systematically tried to enter institutions of higher education. They began by auditing classes at the university and Medical-Surgical Academy in St. Petersburg. By 1862 they had some hope that these institutions would open their doors to women as officially enrolled degree candidates. (This, by the way, would have made Russia the first country in all of Europe to admit women students.) When the tsarist government accurately saw a close connection between the women's aspirations and the nihilist philosophy, and barred women from higher education, the women were forced to seek university degrees abroad.

To a large extent, it was Russian women of the generation of the sixties who opened up higher education to women in continental Europe. Nadezhda Suslova was the first officially enrolled woman student in Zurich, and Sofia Kovalevskaia and Iulia Lermontova were the first women at Heidelberg University. Geneva, Bern, and Paris also had large numbers of Russian women among their first women students. The women were committed as much to social reform as they were to their studies. They rapidly acquired the same reputation as progressive, non-traditional, politically and socially aware individuals as had their male counterparts when they had been sent abroad by the Russian government several years earlier.

It should come as no surprise to learn that the vast majority of Russian women who sought higher education at home and abroad in the 1860s and 1870s chose to specialize in the natural sciences and medicine. In Russia most scientists were sympathetic to nihilism and the women's aspirations, so they welcomed women students. In Western Europe as well, scientists generally proved more willing to accept women students than did their colleagues in the humanities. Most scientists were broad-minded enough to experiment, to let women see for themselves whether they could succeed at the university.

Professors in the humanities, on the other hand, often refused to give women a chance. History had "proved" that women were incapable of higher thought, they said. According to the popular conservative rationale of the time, it would be foolish to allow women to weaken their child-bearing potential by permitting them to engage in anything so injurious as university study. Of course, there were exceptions to these negative views, especially in Russia among younger historians, philologists, and others. But they could rarely persuade their older colleagues to agree to accept women in any capacity. Thus, even women who might have preferred to study the humanities were thrust willy-nilly into the sciences if they wanted any chance at all at university education.

There was another reason for the women's overwhelming choice of the natural sciences and medicine as a subject of study. Women of this first generation were dedicated to proving that they were equal to men in all forms of intellectual endeavor. The sciences were considered the most abstract, difficult, and useful of all branches of human knowledge, so women were bent on showing that they, too, could excel when given the right environment and proper opportunities.

Of the women who grew to adulthood in the tolerant, progressive intelligentsia circles of the 1860s, Sofia Kovalevskaia achieved the greatest professional status and the most enduring scientific fame. In some sense, therefore, she naturally stands apart from her lesser-known contemporaries. But Kovalevskaia cannot be divorced from the nihilist milieu. Unless we take account of this crucial aspect of her background, it is impossible to comprehend her character and life in its entirety. Indeed, if Kovalevskaia's connections to the Russian socio-political movements of the 1860s are ignored, then to the modern observer, burdened with popular stereotypes about science and scientists, it seems as if Kovalevskaia were a woman of contradictions: a mathematician on the one hand, and a feminist and social activist on the other. Only in the context of the early nihilist philosophy is it clear that for Kovalevskaia, being at once a scientist and a political progressive and a woman was perfectly natural.

In Western Europe, the idea of a Russian radical woman in any academic field was anathema to most of the intellectual elite. But opposition did not come so much from Kovalevskaia's fellow mathematicians, who, like their Russian counterparts and like mathematicians in any age, were on the whole tolerant of political and cultural diversity. Rather, it came from professors in other fields, university administrators, and educated European society in general.

Kovalevskaia's scientific colleagues tried as much as possible to shield her from the antagonism of anti-feminists and reactionaries. Yet although they were sympathetic to her, even the mathematicians could not fully comprehend Kovalevskaia's background and its effect on her character. Kovalevskaia had a unique and therefore lonely position as the only Russian and the only woman and the only "child of the sixties" involved in the late nineteenth-century Western European mathematical world. It could not have been easy for her, living abroad in a non-Russian speaking, relatively unpolitical environment.

Despite these difficulties, Kovalevskaia's life in the mathematical and intellectual community of Western Europe was exciting and satisfying. She entered mathematics at a time when stimulating new directions in research were being explored. She studied with Karl Weierstrass, who was one of the foremost mathematicians in the world and is often called "the father of modern [mathematical] analysis."

In her mature years, Kovalevskaia participated fully in the European scientific community. She associated on equal terms with Weierstrass, Charles Hermite, Emile Picard, Henri Poincaré, and other outstanding mathematicians of her time. Kovalevskaia actively contributed to the "mathematical culture" of late nineteenth-century Europe. Because of her early training and scientific background, she had some feeling for practical modes of explanation and down-to-earth exposition, even in the most abstract areas of mathematics. She was therefore able to bring together two mathematical traditions—the concrete approach to theoretical problems favored by mathematicians in her native Russia, and the more abstract "analysis for analysis' sake" tendency of Weierstrass and his school.

Mathematics is one of the most international of all fields of knowledge. Mathematicians were among the first to organize international congresses, and even during the early period drew participants from many parts of the world. Yet mathematics has a local flavor as well. Different "schools" of mathematics develop under the guidance of different masters. The "school"—generally one or a few prominent mathematicians, their junior colleagues and former graduate students—uses specific techniques to solve certain kinds of problems, and occasionally looks askance at other schools and their techniques.

Sometimes mathematical schools form along national lines. In the second half of the nineteenth century, among the main tendencies were the French, German, and Russian schools. Kovalevskaia, with her varied background, excellent expository abilities, and cordial relations with the mathematicians of these three nations, was a perfect candidate to transmit mathematical ideas from one group to another.

The significance of this facet of Kovalevskaia's mathematical activity should not be underestimated. It is a common misconception among non-scientists that mathematics and the other sciences are solitary enterprises: that a lone researcher plods along in an isolated laboratory or study until suddenly she comes up with a brilliant, totally unexpected solution. The reality is far different, however.

Mathematics, like all the natural sciences, is a cooperative, social endeavor. It demands hard work and individual study, but it also entails much interaction with one's colleagues. Mathematicians have to try out their ideas on their fellows. They need to apply the knowledge and techniques of others to their own problems. For this reason journals, correspondence, seminars, and international conferences play a crucial role in the development of mathematics.

Kovalevskaia was an editor of the journal *Acta Mathematica*. She frequently traveled between Germany, France, Sweden, and Russia, and had contact and correspondence with mathematicians everywhere. She could describe the newest developments in Western mathematics to her Russian colleagues, and in turn publicize their results in the West. The 1870s and 1880s were an especially exciting period for mathematics, and Kovalevskaia was at the center of the mathematical life of her time.

There are always dangers in writing a biography. People show different sides of their personality in different circumstances, and letters, diaries, and memoirs often give contradictory accounts of and explanations for events. The case of Kovalevskaia is particularly complicated. She lived in many different worlds, and associated with widely disparate groups of people. None of her contemporaries knew her in all of her aspects. For this reason a collection of reminiscences by a friend or relative, a set of letters to one correspondent, or one of Kovalevskaia's autobiographical writings only illuminates one or two facets of her life. In order to obtain an accurate portrait, the various kinds of information from sources in Russia, Sweden, Germany, and France must be blended together in one story.

This biography attempts to draw together all the elements of Kovalevskaia's life and work into a coherent, well-documented whole. This is the first time that archival sources in both the Soviet Union and Sweden have been used in compiling one account. In particular, the entire collection of the Institut Mittag-Leffler of the Royal Swedish Academy of Sciences was used.

My special thanks go to the Institut Mittag-Leffler, the director Professor Lennart Carleson and the staff, especially Muff

Göransson and Barbro Bjornberg. Their hospitality and kindness in giving me unlimited access to their archives are greatly
appreciated.

Research for the dissertation from which this book comes was
financed in part by grants from the International Research &
Exchanges Board (IREX), the Soviet Ministry of Higher and
Specialized Secondary Education, and the Fulbright-Hays
Commission.

My thanks are also due to Walter Kaufmann-Bühler of
Springer-Verlag. To him belongs the credit (or blame!) for first
encouraging me to work on a biography of Kovalevskaia.

Research was undertaken in various libraries of the United
States and Europe. The staffs of all were helpful, but I would
especially like to thank the Interlibrary Loan Office and Laura
Larsson of the University of Washington, and the librarians of
the Biblioteka Akademii Nauk (BAN) in Leningrad, USSR.
Some of my happiest library hours were spent in BAN, and if
only they could go easy on the *provetrovaniia* (thrice daily room
airings, even in sub-zero weather), the place would be perfect!

My graduate adviser, Professor Norman Naimark, bore with
me for seven years, for which I am grateful. His comments on
the manuscript were extremely helpful, and he has always supported and encouraged my work.

Neal Koblitz and Louise McReynolds read drafts of the
manuscript, and gave invaluable suggestions for revisions.
Roger Cooke read the first draft, and sent me useful information from his examination of the Mittag-Leffler archives. André
Weil read the draft, and I am grateful for his helpful and supportive remarks.

Loren Graham, Deborah Hughes-Hallett, Kenneth Manning
and Richard Stites helped me very much with advice and encouragement on publication. Mutiara Buys, Thomas Glick,
Loren Graham, Diana Long Hall, Deborah Hughes-Hallett,
Natalia Ivanovna Iakovkina, Mark Kac, Walter Kaufmann-
Bühler, Kenneth Manning, Carl Offner, Fritz Ringer, and Eric
Weitz, also read all or part of the manuscript. I thank everyone
for his/her comments. Naturally, any errors are my responsibility.

I would like to add personal thanks to my friends, the
women of Obshchezhitie No. 2 in Leningrad. They took care of

me when I was sick, fed me when I came home late from the library, and in general made my months in Leningrad a delight.

The Library of Congress system of transliteration is used, but soft signs have been dropped from the names of people and places, as have some double i's, as in Mariia, Iuliia. Dates are new style, except for dates in references to Russian newspapers and letters, where changing the date would make the source difficult to locate.

Seattle, Washington
9 June 1983

A CONVERGENCE OF LIVES

INTRODUCTION

On September 29, 1874, the Korvin-Krukovskii family gathered at their country estate, Palibino, to celebrate the feast day of Saint Sofia, which was the name day of the family's younger daughter. The large stone house, with its three-story clock tower and sprawling additions, was full of guests. Relatives from as far away as St. Petersburg, and "neighbors" from the surrounding Nevel district (in the old Vitebsk region, near the border with Lithuania and Belorussia) had been arriving for several days.

In Russia, "that country of great distances," a neighbor was anyone who could "leave home in the morning in a good troika and, [by] diligently whipping up the horses, reach another landowner's home earlier than nightfall."[1] Since feast days in the Korvin-Krukovskii household were traditionally celebrated in a grand manner, no one wished to miss out on the revels, so everyone who could claim the title of "neighbor" was there.

The festivities, as usual, included a play put on by the younger members of the family. The Korvin-Krukovskiis were great lovers of amateur theatricals, and had their own small theater room, complete with raised stage. This time, the play was a French comedy translated for the occasion, and everyone agreed that the actors, even the least experienced of them, performed their parts creditably.

After the play, General Vasilii Vasilevich opened the ball.

---

[1]S. V. Kovalevskaia, *Vospominaniia Povesti* (Moscow: Nauka, 1974), p. 342; hereafter cited as *VP 1974*.

Dancing continued until two or three in the morning, at which time the guests sat down to an elaborate meal. When it came time for dessert, the servants distributed glasses of champagne. Suddenly, the venerable old family tutor, Iosif Ignatevich Malevich, who had retired to a cottage on the Palibino estate, got up to make a speech. "Gentlemen and ladies!" he began,

> We are celebrating today the name day of Sofia Vasilevna Kovalevskaia. Each of us has given her our best wishes . . . but I consider it appropriate to invite you all, dear people, to congratulate her as well on a triumph of feminine achievement; a triumph which has been awarded with the degree of doctor of philosophy.
>
> The woman question, which was brought to the attention of the public in the last decade, divided society into two opposing camps. But it also gave a strong stimulus to many energetic women, and directed them toward independent work in science. The last decade has given Russia many female professionals, who are useful to their fellow citizens in a variety of scientific areas. However, when we probe into the cause of all this activity, it is impossible not to notice that these women's efforts were not entirely disinterested; they did not originate only in the simple love of science for the sake of science itself.
>
> But then a young woman appeared, firm-willed and decisive, determined to pursue her most praiseworthy but extremely difficult goal. She devoted herself to one of the most challenging branches of science, and worked indefatigably in the area of pure mathematics. She married an enlightened man who fully shared her opinions, and did not in the least prevent her from moving forward. She abandoned the pleasures of the world, she sacrificed the best years of a woman's life, she ignored all fatigue, and with rare energy studied her subject in one of the best German universities. But her brilliant successes, achieved in Heidelberg in the course of several semesters, convinced her that her full potential could not be realized there. She moved to the center of German scholarship—to Berlin—and was drawn to Weierstrass, a luminary of science, one of the most famous professors of Europe. She astounded him with her knowledge, and he met with her often to give her valuable advice and instruction. Thus, in the course of five years, she attained the highest academic degree, that degree which in the mathematical sciences is given to very few men.

Now possibly, many of you could ask me the question which has already been asked me more than once: what is the use of studying pure rather than applied mathematics? What is the advantage, especially for a woman? To such questions I answer categorically: Did Newton, Herschel [a prominent British astronomer] and other luminaries of the intellectual world . . . think at any time of the attainment of material gain from their science, from their great discoveries? No, that would have been unworthy of that sphere in which their genius soared, unworthy of science, of which they were high priests. They worked disinterestedly, without remuneration, they worked only for science itself, which was their alpha and omega. In like manner worked, and I hope will work, the respected Sofia Vasilevna. . . .

I salute you, Sofia Vasilevna! You stand on a high pedestal in the ranks of scholars! I salute you also in the name of our small circle, gathered today to celebrate the day of your patron saint! I salute you in the name of our native land, as the first Russian woman to attain the highest academic degree in one of the most difficult areas of science!

Gentlemen, I propose a toast to the health of the first Russian woman scholar, Sofia Vasilevna Kovalevskaia.[2]

Loud cheers sounded in the room. The guests raised their glasses and drank to Sofia's health. Malevich hurried to present a bouquet of flowers to her, but before he could do so he was stopped by his former pupil's counter toast: "Gentlemen, let us drink to the health of my first teacher!" Everyone smiled, and rushed to clink glasses with Malevich and Kovalevskaia, and in the confusion both the flowers and the little speech Malevich had meant to give while presenting them were forgotten.

But perhaps that was just as well, for Sofia had surely had enough of speechmaking. During Malevich's somewhat ponderous presentation, which had come as a complete surprise to her, she had cast down her eyes, blushed, looked displeased, and finally threw herself into her mother's arms, before quieting down to listen to the rest of her former tutor's elaborate

---

[2]Iosif Ignatevich Malevich, "Sofia Vasilevna Kovalevskaia," *Russkaia starina*, 68, No. 12 (1890), pp. 648–51. The translation of Malevich's speech has been edited greatly; Malevich is often pompous to the point of incomprehensibility.

speech.[3] Although she was undoubtedly touched by the attention and honor paid to her in this grand celebration, some of Malevich's comments must have grated on her nerves. For Sofia Kovalevskaia was a true "daughter of the 1860s." In spite of her unusual love for and ability in mathematics, she would never have said that she had devoted her whole being solely and completely to the pursuit of science for its own sake. Nor would she have made a distinction, as Malevich did, between herself and those other Russian women who had struggled so courageously for the right to higher education and a profession. Indeed, several of those women—Nadezhda Suslova, Anna Evreinova, Maria Bokova-Sechenova, and Iulia Lermontova, to name a few—were friends of Kovalevskaia, influenced her development, and helped her at various stages of her career.[4]

Mathematics was an integral part of Sofia's life, and there were times when she did immerse herself completely in her mathematical work. On the other hand, there were periods (one of them lasting over five years) when she occupied herself with just about everything but mathematics. Although her attention had been drawn to the subject when she was about eight, at the age of eighteen she still had not quite decided whether she wanted to make that her life's work. In fact, she sometimes asked herself the same questions Malevich said were asked of him. Namely, what is the good of studying pure mathematics, which seems totally useless from the practical point of view? Would it not be better (that is, more useful) to keep purely abstract studies for one's leisure hours and devote one's life to something more vitally necessary to society?

[3]Malevich, p. 651.

[4]Nadezhda Prokofevna Suslova (1843–1918) was the first woman in Russia and the second in Europe to receive her medical degree; Anna Mikhailovna Evreinova (1844–1919) was the first woman to receive her doctorate in jurisprudence, and also edited the journal *Severnyi vestnik*; Maria Aleksandrovna Bokova-Sechenova (1839–1929), also a doctor, was an oculist and scientific translator, and is thought by some scholars to be the prototype of Vera Pavlovna in Chernyshevskii's *What is to be Done?*; Iulia Vsevolodovna Lermontova (1846–1919) was the first woman to receive her doctorate in chemistry, and worked closely with the well-known Russian chemist A. M. Butlerov.

In the months immediately before and after her marriage in the fall of 1868, Sofia returned to these questions again and again. For a while, she was determined to become a doctor, and audited anatomy and physiology lectures at the Medical-Surgical Academy in St. Petersburg. But she could not bear to relegate her mathematical studies to an unimportant place in her life. Before her nineteenth birthday Kovalevskaia had decided that she did not have the stomach for medical studies, nor indeed for any "practical activity."[5]

This decision occasionally made her feel more than a little guilty, especially when she was in the presence of those women whom Malevich had described as "not entirely disinterested." These were the women who had taken to scientific studies because they considered that such knowledge would give them the greatest possibility of usefulness to society. They agitated and propagandized for the opening of university-level courses to women. (In the early 1860s, in Russia, as in Europe as a whole, universities were closed to women.) They cut their hair and wore plain dark-colored dresses because they wanted to be valued as people, not as decorative but empty-headed dolls. They considered the Russia of their fathers politically, morally, and socially bankrupt, and many of them dedicated themselves to undermining the autocracy. In short, these were the women who called themselves "nihilists," the term used by Turgenev in *Fathers and Sons* to characterize the younger generation.

Kovalevskaia shared these beliefs, considered herself a true nihilist, and, as will be seen, was sometimes willing to go to extraordinary lengths in support of her convictions. But she did not dedicate her entire life to the radical cause, as did, for example, her older sister Aniuta (who was a member of the Paris Commune and the First International Workingman's Association) or her friend Maria Jankowska-Mendelson, the Polish revolutionary.

Moreover, Sofia had a wide range of interests aside from the radical struggle and even aside from mathematics. From the time she was five years old, she had loved the sound of words.

[5]"Pis'ma S. V. Kovalevskoi 1868 g.," *Golos minuvshego*, No. 4 (1916), Letter No. 17.

In her youth, she thought she would grow up to be a poet. She would make up poems in her head, and recite them while bouncing her ball. In fact, she was almost as well known as a writer as she was as a mathematician. She wrote a long, beautifully composed memoir of her childhood years, two short novels, two plays (in collaboration with her friend Anna Carlotta Leffler), some poetry, several topical essays and remembrances of people she had known, and fragments of numerous literary works which she was unable to complete before her premature death at the age of forty-one.

Kovalevskaia's life, though relatively short, was extremely eventful. She was the first woman in the world in modern times to receive her doctorate in mathematics, and the first woman outside of late Renaissance Italy to hold a chair in the subject (at Stockholm University). An editor of the mathematical journal *Acta Mathematica*, she was the first woman on the board of any scientific journal. In 1888 she won the prestigious Prix Bordin of the French Academy of Sciences with a work considered so important that the prize money was raised from 3000 to 5000 francs. The following year Kovalevskaia was awarded a similar prize by the Swedish Academy of Sciences.

For a long time her existence was largely ignored by official educational circles in Russia, partly because she was a woman, but also because she was a "nihilist." However, her fellow mathematicians agitated effectively, and in 1889 Kovalevskaia was elected a corresponding member of the Imperial Academy of Sciences. She was the first woman to be so honored; in fact, the rules of the Academy had to be changed before a vote could be held regarding her candidacy.

In her own lifetime, Kovalevskaia received a lot of attention in the European popular press. Sometimes, she was even considered to be something of a tourist attraction. During her student years in Heidelberg, according to her friend Iulia Lermontova, "people often stopped in the street to stare at her." Later, in Stockholm, when she went iceskating, schoolboys would point her out as the well known lady professor, and then good-naturedly comment on how poorly she skated.[6]

[6]Iu. V. Lermontova, "Vospominaniia o Sofe Kovalevskoi," Kovalevskaia, *Vospominaniia i Pis'ma* (Moscow: AN SSSR, 1951), p. 383; hereafter referred to

Sofia Kovalevskaia knew and was known by many of the great scientists, writers, and political figures of her day. George Eliot once used her to clinch an argument with Herbert Spencer, who claimed that it was impossible for the female mind to comprehend science.[7] Feminist writers often took her as an example of what women could achieve under favorable circumstances. She was praised by such diverse people as the anarchist Peter Kropotkin, the mathematicians Henri Poincaré, Karl Weierstrass, and Charles Hermite, the populist Peter Lavrov, the playwright Henrik Ibsen, and the Russian Grand Duke Konstantin. She was hated by the writer August Strindberg, and slandered by the chemist R. Wilhelm Bunsen. The British algebraist J. J. Sylvester wrote a sonnet in 1886 naming her the "Muse of the Heavens,"[8] and the sociologist Maksim Kovalevskii, a distant relative of her husband, dedicated one of his books to her. She came into contact with such scientific, cultural and political figures as Dostoevskii, Turgenev, Chekhov, Grieg, Ellen Key, Herzen, Chernyshevskii, Mendeleev, Mechnikov, Timiriazev, Helmholtz, and Darwin.

But the life of Sofia Kovalevskaia is interesting only incidentally because of the people she knew. Primarily, her story is absorbing because of what she herself was—an extremely gifted but in some ways perfectly ordinary woman who fought against the prejudices of her time and sometimes won. It is not my intention to portray Kovalevskaia as a superwoman, and it would be anachronistic to call her a champion of modern women's liberation; she was neither. Rather, she was one of many Russian women who spent their young adulthood in the tolerant, supportive atmosphere of the radical "nihilist" circles of the 1860s, and emerged as independent-minded, dedicated, talented professionals of one sort or another.

as *VIP 1951*; Sofia Vladimirovna Kovalevskaia, "Vospominaniia o Sofe Vasilevne Kovalevskoi," *Pamiati S. V. Kovalevskoi* (Moscow: AN SSSR, 1951), p. 151.

[7]Kovalevskaia, "Vospominaniia o Dzhorzhe Elliote [sic]," *VP 1974*, pp. 238–40.

[8]J. J. Sylvester, letter to S. V. Kovalevskaia, 25 December 1886, Kovalevskaia papers, Institut Mittag-Leffler, Djursholm, Sweden. The sonnet was published in *Nature*, 9 December 1886, p. 132.

Sofia's achievements were impressive, but so were those of her friends—Suslova, Lermontova, Bokova-Sechenova, Evreinova, and others. In short, Kovalevskaia is not the only woman of her time who deserves to have a larger place in histories of science and culture than has yet been accorded to her. What distinguishes her from her friends is the considerable amount of information known about her life, and the truly outstanding quality of her talent. Sofia Kovalevskaia was by far the best scientist to emerge from the Russian women-nihilist circles. In fact, she was probably the greatest woman scientist before Maria Sklodowska Curie.

one

EARLY CHILDHOOD AND FAMILY RELATIONS

Sofia Vasilevna Korvin-Krukovskaia was born in Moscow on January 15, 1850. Her father, Vasilii Vasilevich (1801–1875) was the son of a Russianized Polish landowner of the Pskov region, Vasilii Semenovich Kriukovskoi, and his Russian wife.[1] Vasilii Vasilevich attended the Petersburg Artillery Academy, and served in the army from 1819 to 1858. He was a reasonably well-educated man—he spoke English and French fluently, had a fairly wide acquaintance with mathematics and the natural sciences, and knew some of the leading intellectual figures of his day: the populist P. L. Lavrov, the surgeon and supporter of women's education N. I. Pirogov, the artist F. A. Moller, and the writer and orientalist O. I. Senkovskii.[2] He does not seem to have been at all interested in politics in his

---

[1]Kovalevskaia's maiden name has three forms. Her grandfather used Kriukovskoi, with the accent on the last syllable, as do the natives of the village Polibino (previously Palibino) to this day. Sofia herself used that form as a child—she mentions in her memoirs that as a toddler she was told to remember her name by saying "there is a hook (*kriuk*) on the Kriukovskoi gate" (*VP 1974*, p. 10), and sometimes in later life she signed letters that way. But she was christened Krukovskaia (accent on the second syllable) and her father applied for recognition of the family's descent from the old nobility in that name. After repeated rejections of Vasilii Vasilevich's applications, the Senate finally confirmed his claim in 1858, when he had retired from the artillery with the rank of lieutenant general. Then the family was given the right to refer to itself as Korvin-Krukovskii, the "Korvin" signifying their supposed descent from the Hungarian king Matvei Korvin. (*VP 1974*, pp. 509–10, n. 2.)

[2]*VP 1974*, p. 510, n. 4; Malevich, p. 638.

early life. At a time when many of his fellow junior officers were at least peripherally involved in the plotting and secret societies leading to the Decembrist uprising, he steered clear of any such entanglements. His stock answer to any hints or invitations in that direction was that he would never participate in any illegal organization.[3]

After serving in various parts of the country and rising to the rank of colonel, Vasilii Vasilevich married Elizaveta Fedorovna Shubert (1820–1879) in January 1843. Elizaveta Fedorovna was the daughter of a military topographer, infantry general F. F. Shubert (1789–1865), and the granddaughter of F. I. Shubert (1758–1825), a famous astronomer and member of the Imperial Academy of Sciences. She was a lively and beautiful woman and a talented musician, but her marriage to the much older, stern, traditional Vasilii Vasilevich stifled her.

The couple lived in Moscow, where Colonel Krukovskii was commander of the artillery garrison and later (from 1849) of the Kremlin arsenal. At the end of 1843 their first child, Anna (Aniuta) was born and, as was the custom among the nobility, was immediately given over into the care of wet nurses and foreign *bonnes*. The colonel's duties were not onerous, so he spent a great deal of time at the English Club, gambling into the small hours of the night and losing enormous sums of money.[4] Elizaveta Fedorovna was deprived of the chance of going into the high society which she so loved, for most times her husband was not available to escort her. Her diary, at times a rather pathetic document, records her feelings: "My husband does not allow me to take part in the life of the world to which I belong. . . . He is firm, and I cannot change his mind. I am completely robbed of society." Except for a few solemn dinners at the homes of acquaintances, the only time she saw anyone was when Vasilii Vasilevich entertained at home, or when

[3]S. Ia. Shtraikh, *Sem'ia Kovalevskikh* (Moscow: Sovetskii Pisatel', 1948), p. 107; hereafter cited as Shtraikh, *Sem'ia*.

[4]Gambling at the English Club was the weakness of many of the old Moscow aristocrats. See, for example, Peter Kropotkin, *Memoirs of a Revolutionist* (New York: Grove, 1968), p. 31.

visitors came calling. On those occasions Elizaveta Fedorovna would become excited:

> As a reward for that dinner [a boring banquet to which she had been subjected] I spent a delightful evening with Pirogov, who came later. What absorbing conversations, what interesting stories! There is no subject which we did not touch upon in our meeting: religion, love, the family, etc. We left nothing out. Pirogov amazed me with the novelty of his views.[5]

And so life went on, with Colonel Krukovskii spending his time and his money on card games and gypsy dancers, while Madame Krukovskaia entertained people at home, going out whenever she had the opportunity.[6] Their second daughter Sofia (Sofa, Sonia, or Sonechka, as she was called in her family) was born on January 15, 1850, and a son Fedor (Fedia) was born five years later. In 1852 Vasilii Vasilevich received his generalship, and in 1855 the family traveled with him to his new post in Kaluga, where they remained for three years.

According to Kovalevskaia's *Memories of Childhood*, her early years were not happy. The children were entirely in the care of a Russian *niania* (nanny), except for Aniuta who, being much older, had lessons with a French governess. Nanny was kind, loving, and carelessly slapdash in her ways—the perfect nurse for young children. But without realizing what she was doing, Nanny instilled in Sofa the conviction that she was the least loved of the three children. This conviction was reflected in the grown-up Sofa's *Memories of Childhood*.

Kovalevskaia relates how sometimes a maid would come into the nursery when there were visitors, and order Nanny to dress Fedia in some special outfit because the mistress wanted to show him off to the guests. (Aniuta would already be with the adult company.) Instead of simply carrying out the order

---

[5]Both quoted in Shtraikh, *Sem'ia*, p. 108. Later, Pirogov was a surgeon in the Crimean War and was impressed by the abilities of his female nurses. He initiated discussions about women's capabilities in a series of articles advocating women's higher education and other reforms. See essays by D. Atkinson and R. Stites in *Women in Russia* (Stanford: Stanford University Press, 1977), pp. 3–62.

[6]Shtraikh, *Sem'ia*, p. 108.

without any fuss, Nanny would first stop to cuddle and comfort Sofa for having been ignored, thereby magnifying the incident in the little girl's eyes. It was quite common for gentry parents to show preference for their sons, especially when the son's birth had been preceded by that of two daughters, as in this case. Sofa was Nanny's favorite, and consequently the nurse imagined and/or exaggerated parental slights and insults to her darling where, most likely, none were intended.

When Nanny entertained other servants in the nursery at night, Sofa would often hear herself being discussed:

'Wasn't I the one who had to bring her up practically all by myself? The rest of them didn't want to be bothered with her. When we had our Anyutochka, now, her Papenka and her Mamenka and Grandpapa and all her aunties simply adored her, because she was the first. . . . But things were altogether different with Sonechka.'

At this point in the story, so often repeated, Nanny always lowers her voice mysteriously. This, it goes without saying, makes me listen even harder.

'She was born at the wrong time, my little dove, and that's the truth of it!' Nanny says in a half whisper. 'The master, practically on the day before she was born, he goes and gambles his money away at the English Club . . . drops it all . . . they had to go and pawn the mistress's diamonds! . . . the master and the mistress, both of them wanted a son so much. . . . The mistress was so upset, she didn't even want to look at her. It was only later, when Fedya came, that he consoled them.' . . .

Because of stories like these I early formed the conviction that I was not loved, and this belief affected my entire personality. I grew even shyer and more withdrawn.[7]

As a result of her conviction that she was unloved, little Sofa was awkward, clumsy, and sulky with everyone but her Nanny. Elizaveta Fedorovna visited the nursery rarely, but when she did, Aniuta and Fedia would run to her, and hug and kiss her. Usually, Sofa would sit in a corner, too shy to move.

---

[7]Sofya Kovalevskaya, *A Russian Childhood* (Translation of *Vospominaniia detstva* (Memories of Childhood) and introduction by Beatrice Stillman, New York: Springer, 1978), pp. 55–6; hereafter cited as Kovalevskaya, *Russian Childhood*.

Only occasionally would she summon up the courage to embrace her mother; and somehow, these advances on her part inevitably led to disaster. In her self-consciousness, Sofa would end by unthinkingly hurting Elizaveta Fedorovna, or by tearing or soiling the party dress she had come to the nursery to show to her children. The younger girl always felt herself to be shown up in a particularly bad light because such accidents never happened to beautiful, graceful Aniuta.

Nanny's gossip had a further effect. Whenever Sofa was presented to visitors, she would stand stiffly, cling to her nurse's skirts, and glower dreadfully at the assembled guests. "No matter how Nanny tried, I remained stubbornly silent and merely glared at them all, scared and resentful like some little badgered animal," until her mother would finally lose patience and tell the nurse to take her away.[8]

Sofia was discouraged from playing with other children when she was very young, so that when she did get the opportunity to consort with girls of her own age, she was as shy and withdrawn with them as she was with adults. As always, her only refuge was Nanny, who would tell stories to soothe her. But Sofa was an imaginative child, and the stories her nurse told her, with the best of intentions, were not conducive to a calm night's sleep. In fact, she remembered them with horror many years later. In an 1872 letter to her husband's brother Aleksander, she inquired rather anxiously what stories were being told to his daughter: "Do you tell her many tales and does she believe in witches and twelve-headed snakes?"[9]

Five-year-old Sofa began to acquire various fears and dreads. She was terrified, not of darkness itself, but of the shadows associated with the coming of night. She had a horror of large houses with empty spaces where the windows should be, and she could not bear to see a broken doll, or a person subject to fits, or any kind of natural freak.

Some of these phobias remained with her all her life. Her brother Fedia recalled that in later years, she developed the same almost insane terror toward cats as she had had in her

[8]*Ibid.*, p. 56.

[9]*VP 1974*, p. 14; "V. O. i A. O. Kovalevskie," *Nauchnoe nasledstvo* (Moscow-Leningrad: AN SSSR, 1948), I, letter No. 50.

childhood toward broken dolls. Her friend Iulia Lermontova recorded that Sofa's sleep during her student years in Heidelberg and Berlin was beset by fantastic nightmares, some of which had their roots in the stories Nanny had told her as a child.[10]

Fortunately for Sofia's mental health, Nanny's loving but psychologically harmful rule did not continue past her charge's early childhood. As Kovalevskaia admits in her memoirs, "I was on the way to turning into a nervous, sickly child, but soon, however, my whole environment changed, and all previous circumstances came to an end."[11]

The late 1850s, after the humiliating end of the Crimean War, were a time of upheaval in tsarist Russia. Stories of bribery, corruption, and gross mismanagement at the front and in St. Petersburg were circulating everywhere, to the discomfiture of the government and the cynical satisfaction of liberal elements in society. The situation in the countryside was worsening, and the fear of peasant rebellions was increasing. Alexander II and his advisors became convinced of the necessity of eliminating serfdom from above before large scale peasant uprisings did away with the institution from below.

Rumors of the impending emancipation gladdened the hearts of those who had been agitating for change (insofar as they could within the restraints imposed by press censorship and the tsarist secret police). But the rumors upset the many conservative, apolitical, or even mildly liberal landlords who rarely visited their estates because of their positions in government service or their dislike of country life. These landlords, who were dependent upon the revenues from their holdings, began to worry about what would become of their wealth, which was in large part measured by the number of serfs they possessed. Some of the more prudent of them decided to leave government service to attend to their lands. General Korvin-Krukovskii was among their number. In the winter of 1858 he announced to his wife that it was their duty, in those troubled

[10]F. V. Korvin-Krukovskii, "Vospominaniia o sestre," *VIP 1951*, p. 369; Lermontova, p. 386.

[11]*VP 1974*, p. 15.

times, to be at their estate. Elizaveta Fedorovna was unenthusiastic, since she had never lived in the country, but she had no choice but to go along.

The estate of Palibino was a large one, even after part of it had been sold off to meet Vasilii Vasilevich's gambling and other debts.[12] It contained densely forested areas, herds of cows and sheep, a dairy farm, a vodka distillery, and various orchards and gardens. The house was big and rambling, with two wide-winged additions which "seemed to take the guest into their embrace," as one early visitor nostalgically remarked years afterward.[13]

General Korvin-Krukovskii immediately set about the task of learning how to manage his estates, but his enthusiasm did not last long. Soon elected Marshal of the Nobility for his district and then for the region, he found his time increasingly taken up with administrative affairs related to his office.

For the first few months, Aniuta and Sofa ran wild in their new home. They explored the fields, the romantic three-story clock tower, the orchards, gardens, and bramble-bordered country roads. Nanny, of course, would deny them nothing, and the French governess was ineffective at controlling the high-spirited, spoiled Aniuta. Things went on so for a while, with Vasilii Vasilevich ensconced in his study, secure in his misconception that his daughters were being transformed into proper young ladies under the watchful eye of their governess.

But one day he was abruptly undeceived. The girls went on another of their usual exploratory journeys, and were gone all day. When they were finally found, it was discovered that they had gotten hold of some inedible berries and were sick for several days. Much to his indignation, on looking into the matter further, the general learned that Sofa and Aniuta were virtually unsupervised. The girls were both undisciplined, and Aniuta, who at fourteen should have been at least a little

[12]Shtraikh, *Sem'ia*, p. 109.

[13]M. I. Semevskii, "Putevye ocherki, zametki i nabroski," *Russkaia starina*, 68, No. 12 (1890), p. 713. Semevskii, the editor of *Russkaia starina*, was remembering his visit to Palibino in the winter of 1862–63. On that visit he had offered for Aniuta and been rejected by her father. But that he retained pleasant memories of the Palibino family, even including Vasilii Vasilevich, is clear from his frequent mentions of them in his journal.

educated, knew almost nothing. Orders flew with lightning speed: the French governess was dismissed, Nanny was taken out of the nursery and given a job in the linen room, and two new faces appeared on the scene—a Polish tutor, Iosif Ignatevich Malevich, and an English governess, Margarita Frantsevna Smith.[14]

When Kovalevskaia set down her recollections of the period of Miss Smith's domination in *Memories of Childhood*, she tended to emphasize the unhappy aspects of her situation. She recalls six years of being separated from the rest of the family by the stern, unyielding governess, who was jealous of the influence Aniuta had over her charge. Kovalevskaia paints a heartrending picture of little Sonechka, deprived of all affection, ignored by her mother, brother, and sister when she came to them in the evenings, prevented from playing with the village children, and forbidden to read anything other than a small number of approved children's books.[15]

It cannot be denied that at times, especially before the family moved to Palibino, Sofia felt herself to be unloved. However, the adult Kovalevskaia sometimes indulged in a tendency toward self-dramatization, and her *Memories of Childhood* are highly colored by that trait. From the observations of other people, and even from many of Sofia's own comments, one gets the impression that her childhood at Palibino was not so uniformly gloomy as she often liked to portray it.

Country pleasures were simple but extremely exciting for a child who had never experienced them before. *Memories of Childhood* recounts visits, and harvests, and wolves to be listened to on cold winter nights. There were elaborate name day festivities with amateur theatricals and fireworks displays. Palibino was bordered on one side by an immense forest which was a constant source of recreation, edible delicacies, and youthful fantasies. In *Memories of Childhood*, Kovalevskaia describes, with all her remembered delight, the large expeditions into the forest to search for berries, mushrooms, and nuts. She recalls also her childhood fascination with the estate's forester,

[14]*VP 1974*, pp. 27–8.

[15]*Ibid.*, pp. 28–36.

an old Belorussian peasant reputed to be an Old Believer (a member of a stern religious group known for piety and simplicity). "Uncle Iakov," as he was called, told absorbing tales of trees and animals, and brought the manor children such excellent presents as a young elk.[16]

Those who knew Kovalevskaia during this period describe her as lively, gay, impulsively affectionate, with a happy, ringing laugh. Far from being unloved, Sofa was her father's favorite and, however she behaved with her mother, exhibited little fear toward the formidable Vasilii Vasilevich. Her brother Fedia describes how she would run to her father to tell him the events of her day. She obviously had no doubts about his willingness to listen to her, and did not scruple to interrupt the general even when he was reading the newspaper.[17]

When Sofa was about six, on her own initiative she embarked on what her brother considered to be "the first instance of her fully independent intellectual activity." According to Fedia, the grownups did not believe in teaching children to read at an early age, and Sofa desperately wanted to learn. She would examine the newspaper for hours, trying to impress the indecipherable symbols on her mind. Then she would catch some adult, perhaps one of her father's unmarried sisters, and ask the woman to tell her what just one letter was. Sometimes she would approach the governess while she was preoccupied teaching Aniuta, and dance around her until finally, to get rid of Sofa, the exasperated woman would tell her the letter. In this way, asking questions when she could not sound out something (Russian has a largely phonetic system of spelling), Sofa soon taught herself to read and came proudly to her father to show off her skill.

At first, the general could not believe his daughter had done so much on her own. He thought someone had coached her to memorize a particular passage in his newspaper as a joke. When the indignant Sofa proved that she could read anything that was set before her, he was proud. This was the type of

[16]Kovalevskaia, "Palibino," *VP 1974*, pp. 323–32, and 542–43, n. 1.

[17]See Malevich, p. 639; Semevskii, p. 714; Korvin-Krukovskii, *passim*. Sofia's Swedish friend Anna Carlotta Leffler, and her tutor, Malevich, both agree that she was Vasilii Vasilevich's favorite child.

intellectual achievement and initiative which Vasilii Vasilevich found much to his taste. Sofa's character resembled his own, and General Korvin-Krukovskii was pleased about this.[18]

But for Sofa's first few years at Palibino, this skill in reading did not avail her much. Miss Smith, who did not believe reading was a necessary accomplishment for a properly brought up young gentlewoman, would only allow Sofa to read approved children's books. She considered fresh air and exercise to be far more important than books, and used to insist that her charge accompany her on a daily English-style "constitutional." These walks, monotonous and boring for the lively young Sofa, took place in the forest until, as Kovalevskaia gleefully recounts, the pair came upon an enormous she-bear with two cubs. After that experience, Miss Smith would either get a footman to accompany them, or walk elsewhere.[19]

In the winter, when the temperature was low or the wind strong, Miss Smith would allow Sofa to miss the walk, ordering her instead to go into the large drawing room and bounce a ball for an hour. This always proved a temptation to the girl, because that room contained the adults' books she was forbidden by her governess to read. Occasionally, the temptation would prove too much for her, especially if someone had left a book or periodical open on one of the tables:

> I struggle with myself for several minutes. I approach some book and at first only look at it; I turn a few pages, read a few sentences and then run with the ball again, as if nothing had happened. But little by little the reading entices me. Seeing that the first attempts have gone well, I forget about the danger and begin to greedily devour one page after another. It doesn't matter what I've come up with . . . I read from the middle with just as much interest and in my imagination I make up the beginning.[20]

If Sofa became too involved in her reading to hear Miss Smith's return, then the governess would punish her for her misbe-

[18]Korvin-Krukovskii, pp. 370–71; Malevich, pp. 639–40.

[19]*VP 1974*, p. 324.

[20]*Ibid.*, p. 32.

havior by sending her to her father, who would stand her in the corner of his study. This was an indignity neither she nor Vasilii Vasilevich much relished.

During the ball-bouncing sessions, Sofa would sometimes compose poetry. At first, she had written down her juvenile literary effusions. But whenever Miss Smith found them, she would declaim them before Aniuta and Fedia, making them sound even more amateurish and awkward than they in fact were. So she began to say her verses to her ball. The rhythm of the bouncing helped to preserve meter in her poetry, and the governess could not accuse her of neglecting her exercise. Malevich, who was sometimes treated to recitations, thought that her work showed great promise; indeed, in her youth, he dreamed of a literary career for her.[21]

In her *Memories of Childhood*, Kovalevskaia says almost nothing about her actual lessons with Malevich, except to mention in several contexts that she was an excellent student. Malevich, of course, has more to add in this area. However, since his memoirs were in some sense published in response to the absence of mentions of him in *Memories of Childhood*, his claims of influence are probably exaggerated. According to him, Miss Smith was a poorly educated woman who had, if anything, a detrimental effect on the education of the two young ladies. The Swiss governess who succeeded her was, however, quite accomplished and taught Sofa French literature in a short time. All other education and exposure to culture was due to him, Malevich.

But, in the first place, it should be noted that Iosif Malevich and Margarita Smith had been antagonistic toward one another since the time they entered the Korvin-Krukovskii household. During the Polish uprising in 1863, the governess would infuriate Malevich by loudly proclaiming that all Poles were liars and deceivers, and that the tsar was too soft with them. In the second place, Kovalevskaia's impressions of the Swiss governess (whom she never calls by name) do not agree with those of Malevich. According to her, the woman basically left her alone, with the exception of two hours a day of language study.

[21]Malevich, pp. 626, 640–41.

In fact, Kovalevskaia claims that the only thing the governess thought about was saving up enough money so that she could retire to Switzerland.[22]

It cannot be denied that the tutor had some influence on Sofa's intellectual development. However, he seems to have been her political rather than scientific mentor. It was he who guided her thoughts during the time of the Polish rebellion, as will be discussed below. But there were other influential personalities in Kovalevskaia's youth who were equally if not more important than Malevich. For example, she was close to two of her uncles who paid particular attention to her. Like many children, she tended to get sudden, violent "crushes" on people: "the moment any of our relatives or friends showed a little more preference for me than for my brother or my sister, I immediately began to feel for that person an emotion verging on adoration."[23]

Sofa's father's older brother, Petr Vasilevich Korvin-Krukovskii, was one of those who evinced a preference for Sofa's company. Uncle Petr was a gentle, childlike man who had suffered much (his wife, whom he adored despite her contemptuous treatment of him, was killed by her own serfs). Now, he derived his only pleasure from reading and expounding upon what he read to whomever would listen. Sofa, with her imaginative, inquiring mind, was the ideal audience. She would sit, wide-eyed, while he gave her somewhat garbled accounts of the newest scientific discoveries, political theories and current events, or explained his pet economic and social schemes for the betterment of mankind.

It was from Uncle Petr that she got her first exposure to mathematical concepts beyond the elementary arithmetic and geometry she was learning with Malevich:

> Although he had never studied mathematics, he had the most profound respect for that branch of learning. From different books he had accumulated some smattering of mathematical knowledge and loved to philosophize about it, and he often

[22]*Ibid, passim;* Kovalevskaia, "Vospominaniia iz vremeni pol'skogo vosstaniia," and "Kuzen Mishel," *VP 1974*, pp. 346, 340.

[23]Kovalevskaya, *Russian Childhood*, p. 111.

reflected aloud in my presence. It was from him, for example, that I heard for the first time about squaring the circle, about the asymptote, toward which a curve approaches constantly without ever reaching it, about many other matters of a similar nature. The meaning of these concepts I naturally could not yet grasp, but they acted on my imagination, instilling in me a reverence for mathematics as an exalted and mysterious science which opens up to its initiates a new world of wonders, inaccessible to ordinary mortals.[24]

Sofia loved Petr Vasilevich dearly, and was always close to him. Her "Autobiographical Story," written only months before she died, again put him in the most important place among those who guided her first interest in mathematics.[25] In addition, his innocent, impractical, idealistically radical social schemes must surely have had some effect on Sofia's political views. In her later life, in conjunction with Anna Carlotta Leffler, she wrote a play espousing a naive form of utopian socialism, of which Uncle Petr would have been proud.

Sofa also formed an emotional attachment to another of her uncles—her mother's younger brother Fedor Fedorovich Shubert (1831–1877), who worked in some bureaucratic capac-

[24]Kovalevskaya, *Russian Childhood*, pp. 121–22.

[25]*VP 1974*, pp. 367–68. It is interesting to note that this essay, written after Kovalevskaia had seen Malevich's memoirs (he had sent them to her to read) contains only a brief mention of him: "It was so long ago that I've now completely forgotten his lessons; they remain with me as a dim recollection. But undoubtedly they influenced me greatly and had an important significance for my development. Malevich taught arithmetic especially well and originally. However I must confess that at first, when I started to study, arithmetic did not particularly interest me. In all probability, thanks to the influence of Uncle Petr Vasilevich, I was more preoccupied with various abstract discussions, for example, of infinity." (*VP 1974*, p. 370.) This was obviously written in rather condescending appeasement of Malevich's slighted self-esteem. Kovalevskaia had persuaded him to omit from his memoirs a passage virulently attacking Miss Smith, various observations on the characters of Aniuta and V. O. Kovalevskii, and comments on her five-year abandonment of mathematics. The editor of *Russkaia starina*, M. I. Semevskii, suppressed a further polemic denigrating the influence of Petr Vasilevich and A. N. Strannoliubskii, Kovalevskaia's first instructor in higher mathematics. (See passages from Malevich's correspondence with Semevskii quoted in *VIP 1951*, pp. 475–76; see also M. I. Semevskii, letter to S. V. Kovalevskaia, 18 September (?) 1890, Kovalevskaia papers, Institut Mittag-Leffler, Djursholm, Sweden.)

ity in the Ministry of War. Shortly after the Korvin-Krukovskii family moved to Palibino, he came on an extended visit, and immediately endeared himself to the nine-year-old Sofa by mistaking her (perhaps intentionally) for her older sister. Sofa was entranced by her uncle, even, as she tells it, a little in love with him. She would stare at him with her huge, wide-opened green eyes, until finally he said they reminded him of gooseberry preserves, "just as big, just as green and sweet."[26]

Uncle Fedor was good-natured, and not long out of the university, so he was perfectly ready to indulge Sofa in the kind of "grown up" conversation she engaged in with Uncle Petr. Every evening after dinner he would sit her down on his knee, and tell her about things he had learned in his lectures—about infusoria, algae, and the formation of coral reefs. She came to treasure these "scientific conversations," as he called them, because of their adult content, and because they allowed her to be alone with her adored uncle.

The meetings abruptly came to an end, however. One day Olia, the daughter of a neighboring landowner, came to spend the day. Although they did not have much in common, the two girls would usually play happily enough together. Sofa had no playmates, and was glad to descend from her lofty intellectual perch and play games with a companion. But this time she met the arrival of Olia with alarm: what would happen after dinner? Olia would not want to hear about science, and Uncle Fedor, to be polite, would probably tell fairy tales or nursery rhymes.

Sofa decided to make a bargain: Olia could direct their play all day, order Sofa to do whatever she wanted, as long as she would leave Sofa and her uncle in peace after dinner. Olia agreed. At the last minute, though, Olia went back on her word, and events transpired as Sofa had feared. Olia whiningly asked to be included in the after-dinner talk. Uncle Fedor agreed, sat the treacherous girl on his knee, and began to tell fairy tales. Sofa was justifiably furious. She had kept her part of the bargain, played silly games with Olia all day, and now she was to be deprived of her scientific conversation! This was too

[26]*VP 1974*, p. 514.

much for Sofa to bear. She bit Olia on the arm, hard enough to draw blood.[27]

Most adults have some such incident of childish jealousy to relate from their youth. Kovalevskaia's incident is unusual because of the cause of the jealousy. Even as a nine-year-old, she says she was interested enough in scientific topics to want to discuss them at every opportunity. She was even willing to submit to her playmate's whims for a whole day so that she could be assured of an hour's serious conversation. The importance Kovalevskaia claims to have attached to scientific discourse at this young age is impressive.

Until now, relatively little has been said about Sofia's relationships with her parents and brother and sister. Her brother Fedia was five years younger than she was, and at the age of twelve was sent away to a boarding school in St. Petersburg. Consequently he only really associated with Kovalevskaia after she returned from studying abroad in 1874. In any case, he does not seem to have had much to do with his two sisters, and is only rarely mentioned in their correspondence. He studied on the physical-mathematical faculty of St. Petersburg University, and then became a bureaucrat in a government ministry. His niece, Kovalevskaia's daughter Sofia Vladimirovna, recalls that she saw him only once before her mother died. He was spoiled and sickly as a child, and when he inherited Palibino in 1875, gambled it away at cards. In his adulthood, he was spoiled by an unmarried sister of his mother, Sofia Fedorovna Shubert, who left him her house in St. Petersburg when she died. In 1897 brother Fedor married "some Polish lady unknown to us," as Sofia Vladimirovna put it, and he had one daughter. He died in 1919.[28]

In the letter to M. I. Semevskii which accompanied her "Autobiographical Story" for *Russkaia starina*, Kovalevskaia

[27]*Ibid.*, pp. 46–8.

[28]Malevich encouraged the general to send Fedia away to school because "the strong and maybe ineradicable stamp of effeminacy lies on him" (Malevich, p. 643); *VIP 1951*, p. 529, n. 369(1); *VP 1974*, p. 511, n. 7; Sofia Vladimirovna Kovalevskaia, "Vospominaniia o materi," *VIP 1951*, pp. 367–68.

nostalgically reminisces about Palibino. She pictures her old environment clearly, and mentions her sister, father, and even Miss Smith. It is interesting to note, however, that she does not mention her mother. Twice in the course of the letter she recalls the older people at Palibino, and both times ignores Elizaveta Fedorovna.[29] In general, Kovalevskaia referred to her mother only on rare occasions; when she did discuss her it was with affection, but in a superficial way.

The truth is that Sofia was never as attached to her mother as she was to her father. Indeed, she had no strong ties to any woman of her mother's generation or older. Nanny was only a partial exception: Sofia loved her, and visited her whenever she returned to Palibino, but the nurse was her social inferior (she had been a serf until the 1861 emancipation) and could not provide any sort of role model for her. Kovalevskaia found her models in and had the best relationships with women of her own age or slightly older: her sister Aniuta, Maria Aleksandrovna Bokova-Sechenova, Nadezhda Prokofevna Suslova, and others. Maria Aleksandrovna (born 1839) seems to have fulfilled the role of spiritual mother to a number of young women of the sixties. Her influence and that of Sofia's other women friends will be discussed below.

It is difficult to determine whether or not Kovalevskaia's relationship with her mother was typical.[30] In an article on the feelings of mothers and daughters among the nobility, the historian Jessica Tovrov quotes Kovalevskaia on how she feared to hug her mother because she often hurt her or spoiled her dress, and comments that Elizaveta Fedorovna "exceeded the prescribed emotional remoteness."[31] But this comment is misleading. After all, Aniuta frequently embraced and petted her mother, and Kovalevskaia gives the impression that it was

[29]Kovalevskaia, "Avtobiograficheskii zapis' " (letter to M. I. Semevskii, 21 May 1890), *VIP 1951*, pp. 137–38.

[30]Little work has been done in the area of family relations in tsarist Russia. There are two articles in *The Family in Imperial Russia* (Urbana: University of Illinois, 1978) of interest in this connection: Jessica Tovrov, "Mother-Child Relationships among the Russian Nobility," pp. 15–43; and Barbara Alpern Engel, "Mothers and Daughters: Family Patterns and the Female Intelligentsia," pp. 44–59.

[31]Tovrov, "Mother-Child Relationships," p. 36.

primarily her own (and Nanny's) fault that Sofa's attempts turned out poorly. Therefore, it is unclear why Tovrov states that Elizaveta Fedorovna did not welcome caresses, and was thus atypical.

The question remains: was the relationship of Sofia and her mother typical? Most likely, the answer is no. Another social historian, Barbara Alpern Engel, states that mothers and daughters in well-off homes had relatively little to do with each other in the daughters' early years, although the mothers kept watch through the intermediary of the governess. But as the girls approached maturity, they came more and more into the orbit of their mothers.[32] This did not happen with Kovalevskaia.

In the days before her family moved to Palibino, little Sofa conceived of her mother as a beautifully dressed, sweet-smelling, young-looking being who visited the nursery to show off her ball dress and say goodnight to her children. Later in her youth she saw Elizaveta Fedorovna as the spoiled pet of her husband:

> My mother, both in her character and her appearance, was one of those women who never grow old. Between her and my father there was a large difference in age, and my father continued to treat her like a child right up to old age. He called her Liza and Lizok, while she always related to him formally, calling him Vasilii Vasilevich. Sometimes he would even reprimand her in front of the children. 'Again you're talking nonsense, Lizochka!' we often heard him say. And Mama was not in the least insulted by this, and if she continued to insist on her own way, then it was only as a spoiled child insists on her own way past all reason.[33]

Kovalevskaia portrays her mother most often as kindly and beautiful, but empty-headed and ineffectual. She says Elizaveta Fedorovna had no idea how to manage servants, and in *Memories of Childhood* recalls several occasions when her mother was outfaced by Nanny or Miss Smith.[34]

[32]Engel, "Mothers and Daughters," pp. 49–50.

[33]*VP 1974*, p. 35.

[34]*Ibid.*, pp. 12, 19, 24, 35.

Kovalevskaia seems to blame Elizaveta Fedorovna slightly for abandoning her children to governesses. It is hard not to share her feelings, although one can sympathize with the mother as well. Young Liza Shubert had hoped for so much from her marriage and was terribly disappointed. In her diary she writes: "And so I am married. My future is full of hope. . . . I have a fascinating husband (January 21, 1843)." But three days later she says: "There is no full, uninterrupted happiness on the face of the earth. One day is enough to convince oneself of this. Yesterday morning I was joyful, lively . . . but today . . . I cried all night."[35]

Many women would have attempted to compensate for their frustrated lives by concentrating all their attention on their children. By contrast, Elizaveta Fedorovna withdrew from household concerns and poured her unhappiness into her diary. There is no question that, as one Soviet historian put it, she "supervised the upbringing of her children poorly."[36]

This withdrawal does not seem to have been typical of her class and period. Many other women of her day, deprived of the satisfaction of higher education and a career and caught in unhappy, stifling marriages, became determined to help their children escape from the age-old pattern. In particular, they devoted their attention to the education of their daughters. Many enlightened mothers strove to provide their daughters with at least one skill—musical training, for example—so that they would have a small chance for independence.[37] Elizaveta Fedorovna, on the other hand, took no interest in her children's upbringing, especially after a few feeble attempts which were quickly repulsed by the governess.

It was Vasilii Vasilevich who, although he considered it to be women's work, finally looked into the education of his children and brought about changes. It was also Vasilii Vasilevich who, in spite of his usually stern-seeming exterior, according to Kovalevskaia was best able to comfort the children when they were sick:

[35]*VIP 1951*, pp. 462–63, n. 44(1).

[36]P. Ia. Polubarinova-Kochina, "Zhizn' i nauchnaia deiatel'nost' S. V. Kovalevskoi," *Pamiati S. V. Kovalevskoi* (Moscow: AN SSSR, 1951), p. 8.

[37]Engel, "Mothers and Daughters," pp. 52–6.

He never permitted himself the least familiarity with us [an exaggeration on her part] except on those occasions when one of us was ill. Then he would change utterly. His fear of losing us made a different person of him. A rare tenderness and gentleness would come into his voice, his way of speaking to us. No one knew better than he how to show affection, how to joke with us. We truly adored him at such times and cherished these memories for a long time afterward.[38]

It was mentioned earlier that Sofia was Vasilii Vasilevich's favorite child. Despite the frequent self-pitying allusions in her memoirs to being unloved as a child, in fact she knew that she was her father's favorite, and in return favored him over her mother. Her *Memories of Childhood* are always more understanding of the dilemmas and approving of the motives of Vasilii Vasilevich than they are of those of Elizaveta Fedorovna.

For example, by the time of the Polish rebellion in 1863 young Sofa was firmly on the side of the Poles against the Russians. Yet she justified her father's ambiguous stance and sympathized with his predicament:

The situation of my father under these conditions [the imposition of military government in the provinces adjoining the Polish territories] was far from enviable. It is true, he took neither direct nor indirect part in the rebellion itself, in the success of which he did not believe. He felt that the uprising was the affair of only one political party, namely, the nobility, while the whole mass of the population (peasants and middle class) was completely indifferent or even hostile to it. His mother was Russian, and he himself was by his own inclinations more than half Russian, especially since he had lived more than half his life in Russia and from his early years served in the tsarist army. All this should have set him above suspicion, but he was after all of Polish ancestry, and that was enough for him to be considered suspicious. His position as Marshal of the Nobility, and therefore, as representative of that now suspect class, was also thought to be dangerous. All his colleagues in the neighboring provinces had fallen as the first sacrifices to the restoration of order. He was the only one who had happily escaped that fate. But his situation was far from tranquil, especially since he firmly

[38]Kovalevskaya, *Russian Childhood*, p. 104.

resolved not to behave toward the government with cowardly docility but to defend the interests of his constituents to the last degree.[39]

Sofia's mother's stand in social situations was more overtly hostile to the Russian suppressors of the Polish rebellion. She at first refused to invite the new governor to her name day party, because she could not bear to receive a man who had been accused of brutality. Nevertheless Sofia did not consider this as an indication of political thought on her mother's part. Instead, she casually dismissed it as the result of tenderness of heart. Sofia seemed to be less tolerant of her mother's mildly hostile stand than she was of her father's decision to be at least superficially polite to the military governor for their region.

In her attitudes toward her mother Kovalevskaia sometimes appears to take her cues from her father. Vasilii Vasilevich was patronizingly affectionate to his wife; so was Sofia to her mother. It would never have occurred to Vasilii Vasilevich that his wife resented his condescension or had a mind of her own; nor does this possibility occur to Sofia, even with the experience of thirty years behind her (*Memories of Childhood* was written in 1888–89).

In actuality, Elizaveta Fedorovna was a gentle, kind, but weak woman whose feelings toward her husband were complex, especially in the first years of her marriage. Her diaries record more resentment toward Vasilii Vasilevich's slights than Sofia or her father realized. On the other hand, the diaries also reveal a woman passionately in love with her husband:

> Finally, I have found a man who deserves every possible sacrifice. . . . I spend a lot of time alone, but solitude enhances the memories of my husband's last kiss and the hope that I will soon be clasped to his bosom again. Today he kissed me more fervently than ever, and my heart is full of [the thought of] it . . .[40]

---

[39] "Vospominaniia iz vremeni," *VP 1974*, p. 344.

[40] Diary passages quoted in S. Ia. Shtraikh, *Sestry Korvin-Krukovskie* (Moscow: Mir, 1933), pp. 9–10.

It seems likely that Elizaveta Fedorovna became disillusioned when she realized that Vasilii Vasilevich's feelings for her were milder than hers for him. Eventually, she resigned herself to being the object of her husband's casual affection rather than his passionate devotion. But her character was such that she could not channel her thwarted emotions into the care of her family and home.

There were times when Sofia tried to understand the reasons for her mother's withdrawal, failure to develop herself as an individual, and failure even to serve as an influence on her daughters' upbringing. Mother and daughter were too dissimilar, however, for Sofia's attempts at understanding to be successful. Sofia was too energetic, too decisive, too like her father to be able to comprehend Elizaveta Fedorovna's weaknesses. From the time she taught herself to read at the age of six, Sofa would decide what she wanted, and go after her goal with determination. It was otherwise with Elizaveta Fedorovna. "She was an excellent musician, sang wonderfully, spoke many languages, and knew German and French literature well," Kovalevskaia remarked in one of the passages intended for the foreign edition of her memoirs. "Besides which, she had many other artistic leanings, but not one of them attracted her enough to demand some kind of sacrifice on her part, or to go against the tastes or comfort of her family."[41] Since Kovalevskaia would sacrifice much in the course of her career, and often go against "the tastes or comfort of her family," it is obvious that Elizaveta Fedorovna and her younger daughter could never be kindred spirits.

Sofia revealed an affinity for the world of learning early in her life. She was encouraged by her father, her tutor, and her uncles, who watched with pride as her intelligence became evident. True, she encountered obstacles in the form of the anti-intellectualism of her governess and the indifference of her mother. But on the whole, circumstances were favorable for the development and broadening of her fledgling mind. Her Uncle

[41]*VP 1974*, p. 512, n. 3.

Petr and her tutor Malevich had prepared the ground well, not only for her scientific study, but also for her maturing political consciousness. As she entered her teens, Sofia was ready to go further—to embark on the study of higher mathematics and the sciences, and to learn more about the social questions of the day.

two

POLITICAL AND SCIENTIFIC DEVELOPMENT; EARLY INFLUENCES

According to Kovalevskaia's *Memories of Childhood*, the strongest influence in her youth was her sister Anna (Aniuta, Aniutochka).[1] The young Sofa adored her, and thought her the most beautiful and most gifted young lady in all the world. Sofa was always happiest when Aniuta allowed her to enter into whatever was occupying her attention at the moment, and saddest when Miss Smith prevented her from going to her sister. Palibino had no close gentry neighbors with girls of her own age, so Aniuta, although almost seven years older than Sofa, was often glad to have the company of her young sister.

When the girls first arrived at their new home, they went exploring together, and played the usual children's games, although according to their brother, both of them had an aversion to dolls. The young Aniuta loved animals, and used to declare that when she grew up and no one could tell her what to do anymore, she would always carry a little pig around with

[1]Anna Vasilevna Korvin-Krukovskaia (1843–1887) studied and engaged in political activity abroad, participated in the Paris Commune, and, together with the French socialist writer André Leo, edited the paper *La Sociale*. She married Victor Jaclard, Communard and acquaintance of Marx and Engels, and had one son by him. (See I. S. Knizhnik-Vetrov, *Russkie deiatel'nitsy pervogo internatsionala i parizhskoi kommuny* (Moscow: Nauka, 1964), and, by the same author, *A. V. Korvin-Krukovskaia (Zhaklar)* (Moscow: Izd. Vsesoiuznogo Obshchestva, 1931).)

her. As for Sofa, she did not much care for larger animals, but she had collections of butterflies, beetles, herbs and other plants.[2]

During the tenure of the French governess, Sofa had often sat in on Aniuta's lessons just to be near her older sister, even though she was considered too young to learn anything. Soon, she even grew to like the classes, and could answer questions better than Aniuta could. After the arrival of Miss Smith, however, the two sisters were no longer so constantly in each other's company.

Aniuta took an instant dislike to Margarita Smith, and the dislike was mutual. Soon after the new governess arrived in 1858, Aniuta, who was fifteen, announced that she was too old to be in the schoolroom. She moved her sleeping quarters to a vacant room near her mother's, and from then on only endured about an hour a day of Miss Smith's company. Elizaveta Fedorovna was not pleased with the change, but agreed for the sake of family peace. Sofa was left alone with the English-woman, who, partly from dislike of Aniuta, but also from a certain smothering love for her young pupil, tried to keep the sisters apart. As Kovalevskaia points out in her memoirs, that was probably a mistake. Forbidden fruit is always more attractive; Sofa became more fascinated with her sister, and sneaked away to be with her whenever possible.[3]

Aniuta's life in the countryside was not happy. From her earliest childhood she had been pretty and charming, and was consequently spoiled. Her parents would talk about her in her presence in front of guests. They bragged about her accomplishments and laughed at her precocious and somewhat impertinent sallies. She was difficult with her governess and rude to the servants, and sat through five years of lessons during the family's time in Moscow and Kaluga without learning much at all.

In the days before the family moved to the countryside, Aniuta was the acknowledged queen of the children's balls to which she was taken. Her father used to tell her jokingly that

[2]F. V. Korvin-Krukovskii, pp. 369, 372.

[3]*VP 1974*, pp. 28, 61.

when she grew up she would marry a prince. Unfortunately, Aniuta came to believe Vasilii Vasilevich's teasing words. But just at the time when she was beginning to look forward to making her debut in Petersburg society, General Korvin-Krukovskii decided to return to his estate. The sudden end to all her childish amusements and dreams of a brilliant marriage came as a shock to her.

Soon after her arrival at Palibino, Aniuta began to beg her father to take her to St. Petersburg for her presentation. Life was passing her by, she claimed, and it was so boring in the country. Vasilii Vasilevich attempted to explain to her that in those difficult times it was the duty of every landowner to live on his estate, but that comforted Aniuta not at all. In her restlessness, she began to take up one amusement after another, feverishly seeking something to occupy all her energy.

For a while, she developed a passion for horseback riding, but as she had no companion but a groom, and no horse but a farmyard slug, Aniuta soon tired of this activity. Then, she started to read every novel she could lay her hands on. (The girls read and spoke both English and French.) Aniuta particularly liked old English romances which depicted the Middle Ages, and she began to live in those times. She appropriated the three-story clock tower, unearthed old carpets and weaponry to decorate its walls, and headed all her letters "Chateau Palibino." She dressed in a simple white costume, braided her hair in two long heavy plaits, and sat at the tower window embroidering the Korvin crest onto canvas.

This "knightly period passed as quickly as it had come" and was followed by a saintly-philosophical stage.[4] After reading a sentimental romance by Bulwer-Lytton, in which Edith Swan-Neck renounces heaven because her dead betrothed has been condemned to hell, Aniuta began to ponder on the emptiness of life. She gave up novels, started reading inspirational literature, and went through the house with a melancholy expression on her face. Instead of yelling at Sofa or Fedia if they asked her for something, as she had done before, Aniuta now immediately gave in, "though with an air of such soul-crushing

[4]*Ibid.*, p. 51.

resignation that my heart pinched and all desire for merriment disappeared."[5]

This stage continued until the seventeenth of September, Elizaveta Fedorovna's name day, approached. As was the custom, amateur theatricals were planned, and Elizaveta tried to convince seventeen-year-old Aniuta to take part. At first, she would not hear of it, but eventually she let herself be persuaded:

> For the first time it fell to Aniuta's lot to take part in the play with all the rights of a grown-up young lady; she was given, of course, the leading role. Rehearsals began, she discovered in herself a surprising aptitude for the stage. And presto, the fear of death, the struggle of her faith with doubts, the terror of the unknown *au delà*—all vanished away. . . .
>
> After Mama's name day she again cried bitterly, but already for a different reason: Father didn't want to surrender to her pleas to send her to drama school—she felt that her calling in life was to become an actress.[6]

Kovalevskaia recounts Aniuta's gyrations with sympathy and humor, but with a touch of mockery as well. Throughout her life she continued to have ambiguous feelings about her older sister—admiration, love, and respect, but also envy and jealousy:

> I would have gone through hell and high water for my sister, but at the same time, in spite of my warmest attachment to her, in the depths of my soul could be found the slightest bit of envy—that special kind of envy which we almost unconsciously cherish toward people very close to us—those whom we admire very much and would like to emulate in everything.[7]

Kovalevskaia's friend Anna Carlotta Leffler points out that this emotion was healthy, because the jealousy Sofia experienced toward her sister was the kind "which strives to emulate its

[5]Kovalevskaya, *Russian Childhood*, p. 143.

[6]*VP 1974*, p. 56.

[7]*Ibid.*, p. 49.

object, not that which belittles and disparages it."[8] Indeed, Aniuta's influence on her younger sister was in most respects positive.

It is interesting to note that in *Memories of Childhood*, when Kovalevskaia first deals with her sister's political development, she describes the genesis of Aniuta's social opinions in the same jocular, slightly mocking style that she uses to portray the equestrian, courtly, saintly, and theatrical stages of Aniuta's youth. But she quickly loses this ironic, amused detachment. It soon becomes evident that the young Sofa depicted in the memoirs became a staunch advocate of the "new ideas" of the 1860s.[9] Moreover, it is clear that the mature Sofia Kovalevskaia who wrote the memoirs retained her identification with the nihilist movement.

Aniuta was first exposed to the ferment of the new ideas by two young men—Mikhail Ivanovich Semevskii, and the local *popovich* (priest's son), Aleksei Filippovich. In 1862–63 Semevskii, at that time a young officer, came frequently to Palibino to visit his old tutor Malevich, and there became acquainted with the nineteen-year-old Anna Korvin-Krukovskaia. Semevskii discussed with her his ideas on freedom and one's duty to serve the people. In addition he probably showed her, or at least described to her, the articles of the realist publicists which were appearing in *Sovremennik* and elsewhere.

Semevskii proposed marriage to Aniuta. She accepted, but Vasilii Vasilevich refused his consent on the grounds that the young man was penniless. Moreover, her father forbade him to

[8]Anna Carlotta Leffler, "Sonya Kovalevsky," in *Sonya Kovalevsky. Her Recollections of Childhood* (New York: Century, 1895), p. 161.

[9]Much has been written on the ideas of the sixties. In novel form, the best sources are Chernyshevskii's *What is to be Done?* and Turgenev's *Fathers and Sons*. Memoirs contain an enormous amount of information—I found those of N. V. and L. P. Shelgunov, M. L. Mikhailov, A. Ia. Panaeva, L. F. Panteleev, P. D. Boborykin and A. M. Skabichevskii particularly interesting. E. N. Vodovozova's *Na zare zhizni* (Moscow: Khudozhestvennaia Literatura, 1964, 2 vol.) contains delightful accounts of herself as a young girl exposed to the circles of the 1860s. Secondary literature is too extensive to enumerate, but Franco Venturi, *Roots of Revolution* (New York: Grosset & Dunlap, 1966); T. A. Bogdanovich, *Liubov' liudei shestidesiatykh godov* (Leningrad: Academia, 1929) are helpful to start on.

visit Palibino anymore. Aniuta does not seem to have been overly upset. In fact, she was probably more distressed by the prospect of not having Semevskii as a companion in conversation than she was at not having him for a husband. In any case, his place, in the intellectual if not the romantic sense, was soon filled by the young Aleksei Filippovich.[10]

Kovalevskaia notes in her memoirs that "from the beginning of the sixties to the beginning of the seventies, all educated sections of Russian society were occupied with only one problem: family disagreements between the old and the young." A surprising feature of these disagreements was that they were primarily over abstract, theoretical questions and matters of principle rather than the usual concrete, material sources of irritation. This was illustrated in the case of the priest and his son.[11]

Father Filipp had always been proud of his boy. He had been a good, obedient child, and had graduated from the seminary with high standing. But suddenly, Aleksei Filippovich turned obstinate. He refused a lucrative parish and the daughter of the late priest who was offered to him with the parish. (It was the custom for the parish to be considered in some sense the dowry of the priest's daughters, and the candidate to a vacancy generally married one of them.)[12] Aleksei announced his intention to study in St. Petersburg instead. As if that were not bad enough:

> Poor Father Filipp grieved over his son's folly. But he might still have felt consoled if only the son had matriculated in the Faculty of Law which, as is well-known, is the most lucrative. Instead of doing that, however, the boy went into the natural sciences. And during his very first holiday he came home and started talking rubbish to the effect that man, allegedly, is descended from monkeys, and that Professor Sechenov had proved, allegedly, that there is no soul, only reflexes. Poor Father Filipp was so taken aback that he grabbed his holy water basin and started to sprinkle his son from it.[13]

[10]Shtraikh, *Sem'ia*, pp. 110–12.

[11]*VP 1974*, p. 57.

[12]See Gregory Freeze, *The Russian Levites: Parish Clergy in the Eighteenth Century* (Cambridge: Harvard University Press, 1977).

[13]Kovalevskaya, *Russian Childhood*, p. 148.

In accordance with his new beliefs about the equality of all men, Aleksei Filippovich attempted to pay a call on Vasilii Vasilevich. The general was so incensed at his presumption that he instructed his major-domo to tell the young *popovich* that he received "petitioners" only in the mornings. This action on the part of General Korvin-Krukovskii brought the generational conflict into his own home with a vengeance.

Aniuta, obviously seeing in Aleksei Filippovich a possible kindred soul, dared to rebuke her father for his "horrible, unworthy" act. Vasilii Vasilevich was mildly surprised at her temerity, but attempted to joke or embarrass her out of her stance. At dinner he told a funny tale about a princess and a stableboy, making them both seem foolish. But Aniuta, who, according to her father's calculations, should have been reduced to humiliation by this story, listened calmly. Furthermore, she began to take every opportunity of meeting with the priest's son, as if to prove on whose side she stood.

At first, the gossip among the servants was that Aniuta and Aleksei Filippovich were meeting for romantic purposes, but Stepan the coachman speedily disabused his fellows of that notion. He had seen them, he said, and it was funny to watch them. Aniuta would say nothing, but would listen intently while the *popovich* talked, waving his hands, occasionally pulling a tattered book from his pocket and reading to emphasize a point, "just as if he were giving her a lesson."[14]

Of course there was nothing romantic in the meetings. As Kovalevskaia points out, the main charm of the priest's son was that he had just come from St. Petersburg and was familiar with all the newest ideas. "More than that, he had even had the happiness to see with his own eyes (admittedly from afar) many of those great men whom the young people of the time revered."[15]

Aniuta was fascinated. Her curiosity had been whetted by her discussions with Semevskii. Now the priest's son could explain to her so much that had been left unelaborated by the

---

[14]*VP 1974*, p. 59.

[15]*Ibid.*, p. 60. In the Swedish edition of *Memories of Childhood*, the names Chernyshevskii, Dobroliubov, and Sleptsov are inserted here.

abrupt departure of her almost-fiancé. Aleksei Filippovich told her about the new admiration for the natural sciences, the philosophy of materialism, the urgent need for reform of the Russian system of government. He discussed with Aniuta the responsibility of every right-thinking young person to become educated, and to bring that education to the Russian peasantry. He told her about the place of women among the "new people"—educated, serious, uninterested in exterior adornment. He brought her copies of the Petersburg "thick journals": *Sovremennik* (The Contemporary) and *Russkoe slovo* (The Russian Word). Once, he even showed her a copy of *Kolokol* (The Bell), Herzen's forbidden paper.

Aniuta did not immediately take as gospel all of the new ideas:

> Many of them disturbed her, seemed to her too extreme; she took a stand against them and argued. But in every case, under the influence of conversations with the *popovich* and after reading the books brought by him, she developed very quickly, and changed not daily, but hourly.
> By the autumn the priest's son had succeeded in quarreling so thoroughly with his father that the latter asked him to leave and not return for the next holidays. But the seed planted by him in Aniuta's mind continued to grow and develop.[16]

Gradually, Aniuta changed from a capricious young lady to a serious young woman. She stopped begging Vasilii Vasilevich to take her to Petersburg to be presented to society, and indeed, started to look with contempt on the idea of balls and parties. She changed her mode of dressing, simplified her hairstyle, and stopped spending her pocket money on finery. She gave the servants' children reading lessons every morning, and would engage in long conversations with the peasant women when she met them on her walks.

Moreover, Aniuta now studied voluntarily. She ordered boxes of serious books from St. Petersburg and paid for them herself out of the money she formerly would have wasted on clothes. She asked Vasilii Vasilevich to allow her to go to the capital to study further. Her father, of course, refused. Like

[16]*VP 1974*, p. 60.

other gentry fathers of the region he had heard rumors of the shameless things that went on in the "nihilist" circles in St. Petersburg. Rumors of a commune where young ladies of good family lived side by side with young men and did their own chores had spread even to Palibino.[17]

The term "nihilist" (not to be confused with more extreme later uses of the word) was popularized in *Fathers and Sons*, where Turgenev used it to describe the younger generation who denied, supposedly, everything. Young people of the sixties adopted the term for their own, and used it interchangeably with "realist," "new man," "new woman," "man/woman/ son/daughter of the sixties." The definition was vague, but the word basically denoted a person who questioned just about everything in traditional tsarist Russia, had great faith in the natural sciences and the power of education, strongly believed in the equality of women, and desired to be of use to the common people in some capacity.[18]

Aniuta looked forward to joining the nihilists. She persisted in her requests to study for far longer than she had ever persisted before in such a situation, but to no avail. Vasilii Vasilevich eventually lost his patience and shouted: "If you yourself don't realize that it is the duty of every proper young lady to live with her parents until she marries, I am not going to waste time arguing with a stupid girl like you!" Aniuta realized that the situation was hopeless, and stopped asking, but relations between her and her father worsened drastically from that day.[19]

The peace of the Palibino household was at an end. Two hostile camps sprang up, and thrashed out the issues of the day at every mealtime. Vasilii Vasilevich and Aniuta barely

[17]Kovalevskaia called this commune "mythical" (*VP 1974*, p. 58), but in fact for a short while there was such an institution—started by the writer and publicist V. A. Sleptsov—on Znamenskaia street in St. Petersburg. For entertaining, fairly positive descriptions of this commune see the memoirs of Panaeva and Vodovozova, who vehemently refute the rumors of immorality. For lurid descriptions of the commune's supposed decadence see N. Uspenskii, *Iz proshlogo* (Moscow: F. Ioganson, 1889).

[18]Vodovozova, vol. II, p. 28ff.; A. M. Skabichevskii, *Literaturnye vospominaniia* (Moscow: Zemlia i Fabrika, 1928), pp. 125ff., 163ff.

[19]*VP 1974*, pp. 60–1.

spoke to one another, but their every word contained gibes and implied insults which the other could accept or ignore as he or she pleased. Miss Smith seized upon Aniuta's "nihilism" and accused her of planning to run away from home with the priest's son to join the commune. She took to calling Aniuta "young noblewoman in the vanguard," which expression, as Kovalevskaia noted, "sounded somehow particularly venomous on her lips."[20]

In the hope, presumably, of catching her in some disgraceful act, Miss Smith started spying on Aniuta's movements. Aniuta, understandably enough resenting Margarita's actions, began to shroud all her activities in secrecy merely for the purpose of annoying the governess. Sofa absorbed enough of the general atmosphere to start quarreling with Miss Smith. Miss Smith, who always threatened to leave when she felt herself insulted, resigned again. This time, in the vain hope that the absence of the governess would restore peace, Elizaveta Fedorovna accepted her resignation.

The departure of Miss Smith in no way lessened the tensions at the dinner table, and soon there was another source of friction. For some time, Sofa had been noticing her sister's increased preoccupation, and wondered at its cause. But in the last few months before her departure Margarita's vigilance had been more complete than ever, so the younger girl had had no opportunity to question Aniuta. After Miss Smith left, Sofa could be with her sister often, and Aniuta was finally moved to confide in her.

The truth was that Aniuta had written two stories, sent them to Fedor and Mikhail Dostoevskii's journal *Epokha*, and they had been accepted! Moreover, Fedor Dostoevskii himself had written a warm letter accepting one of the manuscripts (late summer 1864). In it he encouraged Aniuta to continue her literary endeavors. He also asked her to write to him about herself: "I would be truly happy if you could find it possible to tell me more about yourself—how old you are and in what conditions you live."[21]

[20]*Ibid.*, p. 61.

[21]"I don't have the letter in my possession now, but I read and reread it so frequently in my youth, and it became so impressed on my memory, that I

After Aniuta showed her this letter, and read her the two stories she had written, Sofa's awe of her sister passed all previous bounds.[22] Sofa was familiar with the name of Dostoevskii, since in a moment of weakness the general had consented to subscribe to *Epokha*, and Aniuta and her father frequently argued about Dostoevskii. (Vasilii Vasilevich considered the writer nothing but a journalist and an ex-convict, because he had served time in Siberia for his participation in the mildly radical Petrashevskii discussion circle.)

More important even than the name of Dostoevskii was the fact that Aniuta was now a writer. Sofa had enormous reverence for anyone whose words would appear in print. Besides that, she was honored by the trust Aniuta had shown by confiding in her.

Vasilii Vasilevich knew nothing of the correspondence between his elder daughter and Dostoevskii. Letters were arranged by means of a relative in St. Petersburg (Anna Mikhailovna Evreinova) and the Palibino housekeeper, Domnia Nikitishna Kuzmina. This woman, who was devoted to Aniuta, could usually go through the mailbag before it was brought to General Korvin-Krukovskii. Like many heads of households, he considered it his duty to read the return addresses of all letters sent to his home, and did not scruple to open any he deemed suspicious.

On September 17, Elizaveta Fedorovna's name day, the housekeeper Domnia was so busy with the guests and all the preparations for the ball and banquet, that the mailbag was brought to Vasilii Vasilevich before she could intercept it.

---

think I can present it almost word for word," Kovalevskaia wrote. Indeed, she was able to reproduce Dostoevskii's letter with only five small mistakes. Compare *VP 1974*, p. 65 with F. M. Dostoevskii, *Pis'ma* (Moscow: Gosizdat, 1928), vol. I, p. 574.

[22] Aniuta's stories, "Son" (The Dream) and "Mikhail" (changed from *Poslushnik* [The Novice] on demand of the censor (letter from Dostoevskii to Aniuta, *VIP 1951*, pp. 468–69)) were published in the 1864 No. 8 and 9 issues of *Epokha*, under the pseudonym of Iu. O—v. The journal was a short-lived effort of the Dostoevskii brothers; it closed down in 1865 for financial reasons because, according to one critic, "the journal had already managed to recommend itself so badly that in 1865 it got scarcely 1300 subscribers," which was not enough. (Skabichevskii, *Istoriia noveishei russkoi literatury* [St. Petersburg: Pavlenkov, 1893], p. 163.)

When he saw an insured letter from the journal *Epokha* addressed to the housekeeper, he immediately called her and had her open it in his presence. Unfortunately, the envelope contained not only a letter addressed to Aniuta, which was bad enough, but also payment for the two stories—around three hundred rubles.[23]

For Vasilii Vasilevich, the sight of the money was the last straw. He summarily dismissed the housekeeper from his service, and refused to show himself to the name day guests. Aniuta and her mother were told that he had had some sort of heart attack, and was too ill to see anyone that evening. Kovalevskaia recounts the story of her father's "attack" in a surprisingly straightforward manner. Since she was always so sympathetic to her father, it apparently never occurred to her to wonder whether his attack was not too opportune to be genuine. Certainly it threw the rest of the family off balance; everyone waited anxiously to know what Vasilii Vasilevich would do.

The next day, the general experienced a miraculous recovery from his supposed heart attack, and summoned Aniuta to his study. He berated her soundly for her boldness. He told his daughter that a girl who could enter into a secret correspondence with an unknown man was capable of anything: "Now you are selling your stories, but perhaps there will come a time when you'll sell yourself!"[24]

Vasilii Vasilevich was a man who believed he had total jurisdiction over the behavior and friends of his daughters, and he felt that it was unfeminine to be a writer. Yet he was no domestic tyrant. For a man of his traditional, patriarchal beliefs, he was surprisingly capable of changing his views. Soon, he had Aniuta read "The Dream" aloud at the dinner table. He was visibly affected by the denouement of the story (a young girl, prevented by her upbringing and circumstances from finding

---

[23]*VP 1974*, pp. 66–9. The fact that Dostoevskii managed to pay Aniuta for her stories was in itself indicative of interest on the part of the writer, even leaving aside the personal character of his correspondence with her. His letters from this period are filled with his financial problems—besides the debts of his journal he had responsibility for the family of his late brother and the son of his dead wife. (See vol. I of *Pis'ma*.)

[24]*VP 1974*, p. 70.

happiness with a poor student, dies of grief).[25] From that time the situation at Palibino improved.

As was seen earlier, in connection with his daughters' education, Vasilii Vasilevich was no lover of half measures. Once he had made his peace with the idea of Aniuta's writing, he not only rehired the housekeeper Kuzmina and allowed Aniuta to continue corresponding with Dostoevskii, but he also gave his permission for the writer to call on his daughter that winter, when the Korvin-Krukovskaia women would be in St. Petersburg.

In her *Memories of Childhood*, Kovalevskaia mentions this *volte face* on her father's part casually. She says that such things "often happened in Russian families: the children reeducated the parents."[26] Strange as it might seem, some of the most stern and conservative-appearing gentry fathers of the 1860s and 1870s did turn out to be flexible in their dealings with their nihilist children.[27] Vasilii Vasilevich was a traditionalist, but he was a well-educated, reasonable man. This would not be the last time he would display tact and wisdom in his dealings with his talented, unpredictable daughters.

Elizaveta Fedorovna and her two daughters arrived in St. Petersburg in February 1865. Aniuta immediately wrote to Fedor Mikhailovich Dostoevskii, informing him that she would be "extremely happy" to see him as soon as possible. He came, and, with a few minor setbacks, their acquaintance progressed well. Dostoevskii managed to ingratiate himself with the Shubert aunts (with whom the Palibino ladies were staying) and soon got on good terms with Elizaveta Fedorovna as well. He even paid some attention to the fifteen-year-old Sofa, and Aniuta proudly showed him her sister's poetic attempts. He became a regular visitor to the Shubert apartment, and there

[25]*Ibid.*, p. 71. Kovalevskaia drastically simplifies the plot of Aniuta's story, which is much more interesting than Sofia makes it seem. See Iu. O—v, "Son," *Epokha*, No. 8 (1864), pp. 1–24.

[26]*VP 1974*, p. 71.

[27]See, for example, the revolutionary Olga Liubatovich's description of her relations with her father in "Dalekoe i nedavnee," *Byloe*, No. 5 (1906), pp. 138–40.

discussed his work, his arrest in connection with the Petra-shevskii circle in 1849, and even his epilepsy.

Sofa listened to all his stories with wide-eyed admiration, although she did not understand everything he said. Once, Dostoevskii talked about the drunken rape of a ten-year-old girl, to the indignation of Elizaveta Fedorovna. Sofa did not understand, but guessed by her mother's reaction "that it must be something horrible."[28]

The author could be an entertaining companion, and Sofa was at a susceptible age. Gradually, she developed a crush on the middle-aged Dostoevskii. In a two-page fragment, written in Russian but obviously intended for the Swedish edition of *Memories of Childhood*, Kovalevskaia writes (Sofa's name is changed to "Tania"):

> There is no doubt that if Dostoevskii could have glimpsed into her soul and read her thoughts, guessed even halfway how deep her feeling was toward him, he would have been touched by the boundless ecstasy which she experienced in connection with him. But unfortunately, it was not easy to see all this. In outward appearance, Tania was still entirely childlike. . . . Tania tried to understand Dostoevskii. . . . She thought about him all the time. When she was left alone, she would repeat in her thoughts everything said by him in their last meeting . . .[29]

While Sofa was dreaming about Dostoevskii, the relationship between him and Aniuta was undergoing changes. At first, Aniuta had been deferential toward the writer. But after a disastrous evening at the Shubert home, during which Dostoevskii made a fool of himself before a roomful of guests, Aniuta gradually began to dominate him.[30] She started to miss ap-

---

[28]*VP 1974*, p. 77. This theme was used in a suppressed chapter of *The Possessed* which was not published until after the October Revolution. The chapter appeared as "Ispoved' Stavrogina," in *Byloe*, No. 18 (1922), pp. 227–52.

[29]*VIP 1951*, p. 131.

[30]It has been pointed out that this was probably the model for the scene in *The Idiot* where Myshkin creates an uproar in the Epanchin house. Aniuta is sometimes thought (by E. H. Carr, for example) to be the prototype for Aglaia Epanchina. (Stillman, Introduction to Kovalevskaya, *Russian Childhood*, pp. 40, 45.)

pointments with him and sew while she was talking to him instead of paying him complete attention. She brought up their political disagreements, and contradicted him at every opportunity. Dostoevskii would walk out, saying that it was useless to argue with a nihilist, but the next evening he would return, and he would begin their dispute all over again.

Unfortunately, by the character of their "mating games," Dostoevskii and Aniuta unconsciously encouraged Sofa's hopes that the writer was beginning to return her feelings. Looking back on those weeks in St. Petersburg twenty years later, Kovalevskaia understood that the courting couple had had no thought of her at all. But the fifteen-year-old Sofa truly believed Dostoevskii when, hoping to irritate Aniuta, he would say that her younger sister understood him better, or was more intelligent or even prettier than she was.

When Sofa discovered Dostoevskii in the process of proposing to Aniuta, she was deeply hurt, and offended as well: she felt that her sister and Dostoevskii had made a fool of her. Then, when she learned that Aniuta had actually refused the writer, she simply could not comprehend her sister's motives. Aniuta tried to explain to Sofa that Dostoevskii demanded too much: "His wife would have to devote herself completely, entirely to him, dedicate her whole life to him, think only of him. But I cannot do that. I want to live myself!"[31]

At the time, the young Sofa thought Aniuta's scruples were absurd. Only as she matured would Sofia be able to understand her sister's fears. Ironically, in later life the sisters' roles were reversed. It was Sofia who would grow to value her independence most highly, and Aniuta who would tie herself to a self-centered, demanding man.

[31]*VP 1974*, p. 88. Later Dostoevskii told his second wife, Anna Grigorevna Snitkina, that he had actually been engaged to Aniuta, but "released her from her promise" because "her convictions were diametrically opposed" to his. (Anna Dostoevsky, *Reminiscences* (translated and annotated by Beatrice Stillman) [New York: Liveright, 1975], p. 56.) Stillman says "there is no evidence that Anna Korvin-Krukovskaya was ever actually engaged to Dostoevsky, nor that he broke the engagement because of ideological considerations, although it is true that she was a political radical." (*Reminiscences*, p. 388, n. 17.) In any case, the Dostoevskii family and the Korvin-Krukovskaia sisters remained friends throughout their lives.

The girls returned to Palibino in the spring of 1865. Aniuta, no doubt, thought the country more boring than ever, but she kept up a desultory correspondence with Dostoevskii, and there was some talk of his spending a summer at Palibino.[32] Sofa probably sighed over his letters (she was mentioned in one of them), but by now she had found more interesting things to do than moon over an unattractive, sickly man more than twice her age.

From the time she was about fourteen, Kovalevskaia rapidly matured both intellectually and politically. She had always been a good student, even under the reign of Miss Smith. Early in life, she had developed a great liking for the natural sciences, as evidenced by her collections of insects and plants, her absorption in talk about the newest scientific discoveries, and her fascination with the abstract concepts of mathematics.

There were other indications of her future path. When the Korvin-Krukovskii family moved to Palibino, they redecorated the old house from top to bottom, ordering wallpaper and other household furnishings from St. Petersburg. But it turned out that they had ordered not quite enough, and the nursery was left without new wallpaper. By chance, the paper used to cover the nursery walls in lieu of wallpaper was the lithographed lecture notes from Professor M. V. Ostrogradskii's course on differential and integral calculus, which Vasilii Vasilevich had taken in his youth. Sofa was fascinated:

> I remember how I spent whole hours of my childhood in front of that mysterious wall, trying to make out even a single sentence and find the order in which the pages ought to have followed one another. From long daily contemplation of them, the appearance of many of the formulas burned itself into my memory, and the text itself left a deep imprint in my brain, although at the time I was studying it I could not understand it at all.[33]

---

[32]Dostoevskii, *Pis'ma*, letters No. 240 (out of order—appears in vol. II) and 245, vol. I. Dostoevskii thought his correspondence with Aniuta important enough to complain to the poet A. N. Maikov about the loss òf one of her letters through the negligence of his first wife's ne'er-do-well son. (Letter No. 307, vol. II.)

[33]*VP 1974*, p. 43.

Sofa desperately wanted to understand what was written on her nursery wall, and consequently was disappointed in her first arithmetic lessons, which did not help her to understand Ostrogradskii's symbols. In fact, when her father first asked her how she liked arithmetic, she was not enthusiastic. But apparently Sofa soon convinced herself that in order to get to the point where she could read her wall, she would first have to learn boring things like division and multiplication. She applied herself to her studies, and went through arithmetic and elementary geometry in a relatively short time.[34]

Normally, a young girl would not have been exposed to as much mathematics as Kovalevskaia was, but two circumstances contributed to Sofa's being allowed to continue her studies past elementary arithmetic. In the first place, according to Malevich, mathematics had been Vasilii Vasilevich's favorite subject, and Sofa was his favorite child, so he was pleased at her precocity in this area. For example, when the tutor told him that Sofa had proved something in a different (but correct) way from the one which she had been told to memorize, General Korvin-Krukovskii was proud of his daughter's inventiveness rather than angry at her temerity.[35]

The second circumstance which enabled Sofa to continue her early mathematical studies was connected with the backwardness of her cousin Mishel. In the summer of her fourteenth year, this cousin came to Palibino to take advantage of Malevich's skill as a teacher.[36] Mishel had just failed his exami-

[34]Malevich, pp. 629–30, 639.

[35]*Ibid.*, p. 639–40. Malevich's claim that Vasilii Vasilevich wanted Sofa to study mathematics is apparently not born out by Kovalevskaia in her "Autobiographical Story." There she says that her father first refused to let her have a tutor in higher mathematics, and only gave in after a while. (*VP 1974*, p. 370.) But the contradiction is more apparent than real. Malevich is talking about algebra and geometry—advanced subjects for a young girl but still relatively elementary, while Kovalevskaia is referring to university-level higher mathematics, which Vasilii Vasilevich obviously did not consider suitable study for a woman. In any case, he did, as usual, give in at the end and allow Sofa to study calculus and differential geometry with a tutor.

[36]Mention of the new Swiss governess in connection with this anecdote would seem to indicate that it took place in 1865, when Sofa was fifteen. But this is almost certainly too late. By 1865 Sofa was studying trigonometry and consequently would have had little interest in the elementary algebra and geometry Malevich was teaching to her cousin.

nations for entrance into the seventh grade at the *gymnasium*, and the subject which had caused him the most trouble was mathematics. Part of the problem was that Mishel did not want to prepare himself for the university. He wanted to go to art school, and was therefore being as obstructive as possible. Malevich and Mishel's mother tried everything they could think of to get him to study, and finally hit upon the idea of asking Sofa to join the lessons. They reasoned that if Mishel saw that a mere girl understood the things that he was purporting to find so difficult, he would drop his indifferent stance and try to learn.

For Kovalevskaia, "this decision had a large influence on my fate as well."[37] With her help Malevich finally managed to force the needed algebra and geometry into Mishel's reluctant brain. As for Sofa, she glided through the summer's studies, absorbed each concept as it was presented to her, and eagerly awaited more difficult mathematical topics. She managed to get hold of an advanced algebra text and read it at night, because her governess did not approve of such books outside the confines of daily lessons.[38] Then she began to look around for other mathematical texts.

One of the neighboring estates was owned by Nikolai Nikanorovich Tyrtov, a professor of physics at the Petersburg Naval Academy and a co-founder of free pedagogical courses for women in the capital.[39] On one of his visits, Tyrtov brought a copy of his beginning physics textbook, which Sofa immediately seized upon. Although the book was an elementary one, nevertheless it made use of trigonometric functions, which Sofa had never seen before, and which Malevich was unable to explain to her. She puzzled over the formulas for a long time, and finally decided that a chord would serve her as well as the problematical sine. Once she had reasoned this out, she proceeded to go through the book with little difficulty.

When Tyrtov next visited, Sofa tried to talk with him about his text. He was reluctant, and said he did not see how she

[37]"Kuzen Mishel," *VP 1974*, p. 338.

[38]"Avtobiograficheskii rasskaz," *VP 1974*, p. 370.

[39]For a brief history of these courses, see Liubov Andreevna Vorontsova, *Sofia Kovalevskaia* (Moscow: Molodaia Gvardiia, 1957), pp. 73–4.

could have understood it, since she could not possibly know any trigonometry. Sofa was naturally offended at this disparagement of her mathematical level, and entered into a long explanation of how she had substituted a chord for the sine.[40] Tyrtov became excited, and said that this was the approach used historically in the development of trigonometry. It was "as if she had created that whole branch of science—trigonometry—a second time."[41]

Tyrtov strongly urged Vasilii Vasilevich to let Sofa study higher mathematics with a tutor. Tyrtov was sympathetic to the progressive movements of the 1860s, and believed women were as capable of intellectual achievement as were men. He felt it would be a crime to let Sofa's mathematical abilities go to waste. Eventually, Tyrtov persuaded his friend. General Korvin-Krukovskii agreed to allow Sofa to study trigonometry and calculus.

[40]At small angles in a circle of radius r = 1, a chord and a sine are close in value:

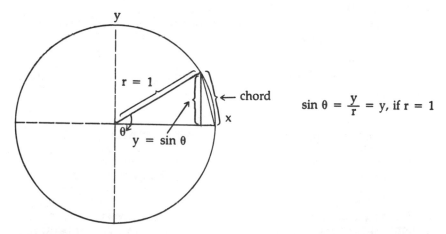

$$\sin \theta = \frac{y}{r} = y, \text{ if } r = 1$$

Tyrtov's book only used very small angles, so Kovalevskaia's method worked.

[41]Korvin-Krukovskii, pp. 371–72. "I am afraid," Korvin-Krukovskii says, "that some readers will find this incident exaggerated. But I can vouch for its accuracy fully, because I was told it on the one hand by my sister herself, and on the other hand it was confirmed by my father and the late N. N. Tyrtov." Kovalevskaia also tells this anecdote, though in less detail, in her "Autobiographical Story."

The incident with Tyrtov, and his praise of her abilities, made Sofa more interested than ever in the natural sciences and mathematics; she studied them as much as she could. When Elizaveta Fedorovna took her daughters to Germany and Switzerland in 1866, sixteen-year-old Sofa bought herself a little microscope and spent hours looking at the cells of plants and insects.[42] Then, in one of her letters home to Palibino she asked her father to bring her mathematics books with him when he, Fedia, and Malevich joined them that winter.[43] Sofa clearly knew where her abilities lay—she only needed the proper environment to develop them to the fullest extent.

During the years when Kovalevskaia was finding herself more and more drawn to the natural sciences and mathematics, she was also maturing politically. Sofia was as consistent in her interest in politics as she was in her scientific interests. In fact, they developed concurrently. The young Sofa's political ideas were naive at first, and retained a certain utopian aspect throughout her life. But the orientation of her views remained fairly constant from the time she was about fifteen until the time of her death.

Kovalevskaia once wrote to her friend, the Polish revolutionary Maria Jankowska-Mendelson: "In my youth I never dreamed of anything as fervently as I dreamed of participating in the Polish rebellion."[44] In 1865, Sofa came close to getting her wish.

[42]Ostensibly, the trip was for Aniuta's health (Malevich, p. 627), but according to Shtraikh the real reason was to separate her from further contact with Dostoevskii. General Korvin-Krukovskii had written him a condescending letter which he hoped would hint Dostoevskii away from any further correspondence with his daughter. (26 January 1866, reproduced in Shtraikh, *Sem'ia*, pp. 124–25.) But the letter failed in its object, so more extreme measures were taken. Aniuta was ordered to write to Dostoevskii with the information that the family would be abroad for the summer, and consequently could not entertain him at Palibino. In fact, they only left for Western Europe in the fall of 1866.

[43]Kovalevskaia to S. Adelung, 14 March 1867, in S. Adelung, "Jugenderinnerungen an Sophie Kowalewsky," *Deutsche Rundschau*, Bd. 89 (1896), pp. 406–07; Malevich, pp. 631, 642.

[44]Maria Mendelson, "Vospominaniia o Sofe Kovalevskoi," *Sovremennyi mir*, No. 2 (1912), p. 182.

The Korvin-Krukovskii estate was situated in a border region with a mixed population. Most landowners were Polish, at least by ancestry, but the peasants were Belorussian or Russian, the main language was Russian, and the principal religion was Russian orthodoxy, although sectarianism was widespread. Since the landowners' children were raised by serfs, they spoke Russian rather than Polish from infancy. Apparently, the parents in the Korvin-Krukovskii circle did not make any effort to see that their children learn Polish along with Russian and French.

Until the time of the Polish uprising, Sofa had not paid much attention to the fact that by birth she was one-quarter Polish. However, after listening to the dinner table arguments between Malevich and Aniuta on the one hand, and Miss Smith on the other, she began to feel sympathy for what she saw as the heroic battle of the Polish people against the tsar. She started to have discussions with Malevich about Polish literature and history, and asked him to teach her the language. The tutor complied, and during the lessons told Sofa stories about the brilliant past of Poland, "our deeply-loved mother."

Miss Smith did not consider newspapers suitable reading matter for a young lady, but Sofa was able to get hold of them by sneaking down to the drawing room at night and reading them by the light of an icon lamp. She was fascinated by reports of the fighting in Warsaw, and dreamed of participating. She had developed one of her crushes on a young neighbor Buinitskii, and when he went off to join the uprising she worshiped him as a hero.

As news of the rebellion became more dismal, Malevich lost his enthusiasm for his "deeply-loved mother," and the Polish lessons and talks on Polish history ceased.[45] But Sofa remem-

---

[45]"Vospominaniia iz vremeni pol'skogo vosstaniia," *VP 1974*, pp. 346–50. This article was originally intended for inclusion in *Memories of Childhood*, but was left out because of its political content, which the censor would not permit. Interestingly enough, "Memoirs from the Time of the Polish Rebellion" has disappeared again from the newest edition of Kovalevskaia's literary works, *Izbrannye proizvedeniia* (Moscow: Sovetskaia Rossiia, 1982). Although this edition is "selected" rather than collected works, the other three variant chapters from *Memories of Childhood* do appear. It will be left to the politically astute reader to guess the reason for this conspicuous omission.

bered Buinitskii. In 1864, news came that his estate had been confiscated, and it was rumored that he had fled abroad, or been shot or exiled to Siberia. Sofa decided that he was in Siberia, and dreamed of killing the military governor of their region, thereby joining Buinitskii in exile. She fantasized about their reunion, and cherished more than ever a poem he had written in her album before he left for Warsaw. After Miss Smith's departure, Sofa devoured accounts of the cruelties committed by the Russian occupiers, and was full of indignation against the Russian troops. Thus, by the time of Elizaveta Fedorovna's name day party in 1865, Sofa was horrified at the idea of entertaining one of their number: the dreaded Colonel Iakovlev.[46]

All went well at the name day celebration. The Polish gentry were stiffly polite to the colonel, and he in his turn tried to show what a cultured, suave gentleman he was. After dinner, the assembled guests began to pay attention to Sofa, and the governess ordered her to bring her album downstairs to show them. Sofa was appalled. The album contained Buinitskii's priceless poem! However, she had no choice but to obey. "The previous night I had slaughtered Colonel Iakovlev in my imagination, but that was a completely different matter than openly to defy my governess!" Unfortunately, the colonel had the impulse to draw something in Sofa's album for her to remember him by, and then invited her to admire it:

> Almost mechanically, I stood and approached him. But when I neared that vandal and saw my poor outraged album in his paws, a feeling of fierce malice took possession of me. Without thinking about what I was doing, I snatched the album from his hands, quickly tore out the page with his drawing and ripped it into little pieces, which I threw on the floor with one word—'*voilà*'.[47]

[46]1865 is the year given by Kovalevskaia, although she goes on to say that she was eleven years old at the time. But Kovalevskaia often tended to lower her age from one to four years in the pages of her memoirs. Therefore, the naming of a specific year, if supported by other evidence, is more reliable than her statement of her age at any given time. In this case, 1863 would have been too early, and 1864 was the year of the name day party at which Aniuta's writing was discovered.

[47]*VP* 1974, p. 376.

According to Kovalevskaia, at first Iakovlev did not know how to react to Sofa's action. On the one hand, she had humiliated him, but on the other hand, he would look even more foolish if he took offense at the act of a fifteen-year-old. He decided to accept the explanation offered him (that Sofa tore up his picture because she was jealous of his drawing skill) and the tense situation passed off rather well. Sofa was sent up to the nursery and regaled with sweets by all the Polish guests who ostensibly came to scold her. Meanwhile, Vasilii Vasilevich purposely lost several hundred rubles to Colonel Iakovlev at cards—"a method frequently used in Russia when you want to give someone 'baksheesh' [a bribe or gratuity] but it is impossible to do so openly."[48]

The governess wanted a more severe punishment for Sofa than mere banishment to her room, but General Korvin-Krukovskii would not allow it. He said it was the grownups' fault for discussing politics in front of her, and that she had been punished enough. Kovalevskaia recalls that he scolded her "in a voice which he tried to make very stern, but I sensed that in the depths of his soul he was not very angry."[49]

As at other times, Vasilii Vasilevich seems to have been secretly pleased at the initiative, determination, and even the political stance displayed by his Sofa. This unobtrusive support of General Korvin-Krukovskii for his younger daughter appears to have been important for Sofa's intellectual development. Interestingly enough, two other great women scientists, the physicist-chemist Maria Sklodowska Curie and the mathematician Emmy Noether, also had loving, close relationships with fathers who proudly encouraged their gifted daughters.

Many people contributed toward influencing Kovalevskaia's early political views—Aniuta, Uncle Petr, Buinitskii, Malevich, and indirectly, Tyrtov, Semevskii and Aleksei Filippovich. Sofa was exposed to progressive ideas from early adolescence. She absorbed her uncle's naive socialism and faith in science, Buinitskii's and to a lesser extent Malevich's hatred of tyranny

[48]*Ibid.*, p. 357.

[49]*Idem.*

and the oppression of one nation by another, Aniuta's and her mentors' egalitarianism and incipient populism. These views, combined with her interest in the natural sciences and her desire for further mathematical instruction, would make Sofia an eager and receptive pupil of the young radicals she would meet in St. Petersburg in the winter of 1867–68.

three

A FICTITIOUS MARRIAGE*

One day in the winter of 1867–68, a servant ushered three young women into the study of a young professor at St. Petersburg University. They sat down solemnly, facing their host. He was nonplussed at this feminine invasion, since he knew them only slightly, but nevertheless he waited politely for them to state their business. After a moment's awkwardness, the eldest asked if he would be willing to "liberate" them all by marrying one of them, accompanying them to a German or Swiss university, and then leaving them there. It did not matter which one of them he chose, she explained, because the other two would be able to go abroad to study under the chaperonage of the "married" one. She made her request seriously and calmly, with no trace of embarrassment. The professor refused, equally composedly, and the four young people shook hands before parting. The three women were Sofia and Aniuta Korvin-Krukovskaia (Aniuta was the spokeswoman), and the sisters' cousin Anna Mikhailovna Evreinova (nicknamed Zhanna).[1] This was their first attempt to arrange a fictitious marriage.

Women in Russia of the 1860s had certain rights which most of their European counterparts had not yet attained.[2] They con-

*Portions of this and the following chapter have appeared in my "Science, Women, and Revolution in Russia," *Science for the People*, 14, No. 4 (1982), pp. 14–8, 34–7.

[1]Leffler, "Sonya Kovalevsky," pp. 163–64. Kovalevskaia later met the professor again and they had a good laugh over the incident.

[2]For information on the status of women in Russia see Dorothy Atkinson,

trolled their own property, for example, and female property owners could vote (through a male proxy) in the municipal and regional elections. Moreover, the Imperial government had forbidden wife-beating in the first half of the nineteenth century. Of course, this did not stamp out the practice, but the law put Russian women in a somewhat better position than were most other European women. In one respect, however, they were entirely dependent on their fathers and husbands. Women were listed on their father's or husband's internal passport, and therefore could not work, study, or even live apart from them without express permission, which traditionally was not forthcoming.

One of the most common ways of circumventing parental authority came to be the "fictitious marriage."[3] A young woman desirous of leaving home to work or study would come to an agreement with a man who would go through the marriage ceremony and then, theoretically at least, leave the woman to pursue her own life. Her father had no further authority over her, and her "husband" was honor-bound to keep their relationship platonic.

Of course, things did not always go according to plan.[4] Moreover, not all young progressives approved of fictitious marriage.[5] Instead, they looked forward to an equal, loving relationship with a member of the opposite sex. But the alterna-

---

"Society and the Sexes in the Russian Past," *Women in Russia* (Stanford: Stanford University Press, 1977), pp. 3–38; Richard Stites, *The Women's Liberation Movement in Russia* (Princeton: Princeton University Press, 1978); E. O. Likhacheva, *Materialy dlia istorii zhenskogo obrazovaniia v Rossii. (1086–1901)* (St. Petersburg: Stasiulevich, 1893–1901, 6 vol.); S. S. Shashkov, *Sobranie sochinenii* (St. Petersburg: Sorokhodov, 1898), vol. I; Marie Zebrikoff, "Russia," in *The Woman Question in Europe*, Theodore Stanton, ed. (New York: Putnam, 1884), pp. 394–401. Stanton's book also gives some good comparative data.

[3]If one had the misfortune already to be married to a tyrant, the only remaining alternative was flight, followed in extreme cases by falsification of papers. Divorce was difficult, and the husband or father had the right, if he chose to exercise it, to enlist police aid to drag back a recalcitrant wife or child.

[4]Elizaveta Vodovozova recounts several tales of fictitious marriages that went awry in her memoirs. (See Vodovozova, vol. II, pp. 223–24.)

[5]See L. P. Shelgunova's comment in her *Vospominaniia* (Moscow: Khudozhestvennaia literatura, 1967), vol. II, p. 119.

tives to fictitious marriage were few. Sofia Korvin-Krukovskaia became one of the first to avail herself of this method.[6]

Both Aniuta and Sofa were ready for rebellion when the Korvin-Krukovskii family came to spend the winter of 1867–68 in St. Petersburg. Aniuta was seething because the general had interfered with her correspondence with Dostoevskii and refused to allow her to leave Palibino to study. Through letters from her cousin Zhanna Evreinova (whose father was a general living outside of Petersburg), Aniuta had learned of many nihilist women who had left their gentry nests to congregate in the capital. They cropped their hair, dressed simply, performed their own chores, conversed on an equal level with the nihilist men, and had a voracious appetite for all learning, especially the natural sciences and medicine.

There were several reasons for this special interest in the sciences. Western scientific and pseudoscientific writers— Darwin, Büchner, Buckle, Moleschott, Spencer—were embraced for the materialism and faith in progress of their world views, which seemed to promise an end to the tyranny of religion and autocracy. There was a general conviction that the spread of knowledge, especially scientific knowledge, would hasten the day of revolution. Moreover, it was believed that science was well on the way to solving all of the world's problems—hunger, poverty, disease. The more people immersed themselves in scientific study, it was thought, the closer mankind came to finally conquering all of its ills.[7]

For the women, there were additional reasons for this interest in the natural sciences. Some chose these fields because they wished to prove that women were capable of successful endeavor in them, or because they specifically wanted to help

[6]S. V. Panteleeva, "Iz Peterburga v Tsiurikh," in L. F. Panteleev, *Vospominaniia* (Moscow: Khudozhestvennaia Literatura, 1958), p. 691.

[7]See Alexander Vucinich, *Science in Russian Culture* (Stanford: Stanford University Press, 1970), Introduction and *passim*. This attitude was to change in the late sixties and early seventies, when conflict arose between acquiring scientific knowledge and working for the revolution. (See J. M. Meijer's excellent work *Knowledge and Revolution* (Assen: Van Gorcum, 1955); Philip Pomper, *Peter Lavrov and the Russian Revolutionary Movement* (Chicago: University of Chicago Press, 1972), p. 94ff. and *passim*.)

poor women who had never seen a doctor or nurse. Not least important, however, was the fact that professors in the natural sciences and medicine, who were often little older than their students and steeped in the new ideas themselves, were usually more willing to accept female pupils than were their colleagues in the humanities.[8]

Cousin Zhanna's news from the capital corroborated and supplemented what the Korvin-Krukovskaia sisters had learned already from Aniuta's erstwhile suitor Semevskii and her friend the priest's son. Sofa and Aniuta became increasingly eager to join in the new, progressive circles which St. Petersburg had to offer. They looked forward to meeting the new women and men of whom Zhanna wrote so admiringly. Sofa was especially eager, because the scientific bent of the nihilists was much to her taste.

Aniuta's plans were not well-formulated, since she had no scientific leanings. She had the vague idea of learning more about life so that she could write convincingly, and she was sincerely interested in the new social and political ideas. The main source of Aniuta's restlessness, however, seems to have been a conviction that life was passing her by. She was twenty-four years old, and she was impatient to be off and doing the great deeds of which she was sure she was capable.

Sofia, on the other hand, was more definite about her goal. She wanted to study mathematics and the natural sciences, and hoped to become a doctor. The year before, she had set up an infirmary for the peasants on her father's estate, where she treated the former serfs of the region. Now, she planned to receive medical training in St. Petersburg, and perhaps practice as a physician in Siberia. There, she would be able to treat political exiles, and study mathematics and the natural sciences in her spare time.[9]

Some writers who have dealt with the relationship between the Korvin-Krukovskaia sisters give most of the initiative in any

[8]See the memoirs of Kropotkin and Vodovozova, my article "Science, Women, and Revolution in Russia," and Chernyshevskii's *What is to be Done?* for more information on this phenomenon.

[9]Kovalevskaia, letter to Aniuta (July 1868), "Pis'ma S. V. Kovalevskoi 1868 g.," *Golos minuvshego*, No. 2 (1916), p. 237.

joint venture to Aniuta. The Soviet writer L. A. Vorontsova calls Sofia "the mirror" of Aniuta. Elsewhere, she describes Sofia as "that shadow of her sister," and says that she only became more firm after she met her future husband, Vladimir Kovalevskii. Beatrice Stillman claims that "it was Anyuta who had shown the courage to fight for independence, but Sofya to whom it had been presented as a gift." Anna Carlotta Leffler remarks that "Sonya kept herself in the shadow of her more brilliant sister. . . . She was so entirely Aniuta's shadow that it was impossible to imagine the one without the other."[10]

These evaluations are unfair to the young Sofia. It is true that Kovalevskaia's own *Memories of Childhood* often portray her young self as moving in the shadow of her older sister. She had worshiped her, thought her superior in all things, and followed Aniuta's changing beliefs from the time she could first talk. The mathematician Elizaveta Litvinova, a friend of Kovalevskaia and the second woman ever to receive her doctorate in mathematics, recalled that Aniuta would harangue her younger sister from the time Sofa was five. She would become exasperated when the little girl's tongue slipped, as it did on occasion, and Sofa said "blanket" (*odeiálo*) instead of "ideal" (*ideál*). According to Litvinova, the girls' mother Elizaveta Fedorovna would sometimes come across her daughters in the following situation: Aniuta would be standing close to her younger sister, declaiming something or other in a fervent voice, and shaking her by the shoulders or gesticulating wildly. Sofa, little more than half the size of her sister, would stare up and try desperately to understand Aniuta's monologues.[11]

That such incidents were part of the girls' existence at Palibino is undoubtedly true, but they by no means convey the whole picture. The previous chapters contain several examples of Sofa's persistence and purposefulness: her success in teaching herself to read, the incident of Colonel Iakovlev and Buinitskii's poem, her ability to persuade her father to allow her to continue her studies even to the level of differential and integral calculus.

[10]Vorontsova, pp. 100, 76, 80; Stillman, Introduction to Kovalevskaya, *Russian Childhood*, p. 10; Leffler, "Sonya Kovalevsky," pp. 161, 163.

[11]E. F. Litvinova, *S. V. Kovalevskaia* (St. Petersburg: Pavlenkov, 1893), p. 16.

Interestingly enough, none of these incidents are mentioned by Kovalevskaia in her *Memories of Childhood*. In general, in her memoirs Kovalevskaia downplays her own determination and intellectual independence. Instead, she focuses on her sister, making Aniuta appear the center of Sofa's life. It must be remembered, however, that *Memories of Childhood* was written in the months following Aniuta's death, and constitute Sofia's memorial to her deceased sister. Consciously or unconsciously, Kovalevskaia appears to have exaggerated the extent of her youthful dependence upon Aniuta. Other sources do not support her picture. Sofa's tutor Malevich, her brother Fedia, and her Aunt Briullova all considered her to be of strong character even in childhood.[12] As she matured, her instinctive deference toward Aniuta became tinged with exasperation. Sofia began to assert herself more frequently, and even tried to direct and help Aniuta.

It is quite possible that without Sofia's influence Aniuta would have been content to mingle in radical circles, envious of those women who had freed themselves, resentful of Vasilii Vasilevich's authority, but nevertheless passively accepting her fate; content, in fact, to dream of freedom in the congenial atmosphere of St. Petersburg. Sofa, on the other hand, became steadily more determined to act, and act soon.

In Petersburg, Sofa took lessons in calculus from A. N. Strannoliubskii (1839–1903), a well-known pedagogue popular among the young people of the sixties for his championship of women's and workers' education.[13] As the Soviet scholar P. Ia. Polubarinova-Kochina emphasizes, the relationship with Strannoliubskii was important for Sofa's further intellectual and political development:

> Strannoliubskii was an excellent teacher and in the future accomplished much for the growth of higher education for women in Russia. . . . He was a fervent partisan of the ideas of the

[12]See Malevich, *passim;* Korvin-Krukovskii, *passim;* letters from Aunt Briullova to Kovalevskaia's cousin Sophie Adelung, published by the latter in "Jugenderinnerungen," pp. 394–425.

[13]See V. E. Prudnikov, "A. N. Strannoliubskii—pedagog i matematik," *Pamiati S. V. Kovalevskoi*, pp. 120–32. Kovalevskaia says that she was fifteen when she first received lessons from Strannoliubskii.

revolutionary democrats—Dobroliubov, Chernyshevskii, Pisarev—and he was a founder of schools for the common people.

Sofa was in ecstasy over Strannoliubskii. He not only taught her mathematics, but also undoubtedly inspired her with high enlightening aspirations.[14]

There can be no doubt that Strannoliubskii took an active interest in Sofa's future and helped to determine her priorities in life. Moreover, through Strannoliubskii Sofa became known to the most serious representatives of the generation of the sixties. Meetings with her tutor and his friends "prepared her to a great extent for that tireless battle for the rights and happiness of women which Sofia Vasilevna was to wage for the rest of her life."[15] With the encouragement Strannoliubskii and his circle gave her, Sofia was ready to look into the possibilities of formal instruction. She wanted to get on with her task of helping the common people, and so needed medical training. As she soon discovered, however, possibilities for formal education were few.

The movement for women's higher education had sprung up as part of the general progressive-egalitarian-democratic movement after the end of the Crimean War.[16] Belief in the power of education, the conviction that everyone, including women, must have the right to receive an education, and a desire eventually to use one's knowledge to help the common people were essential tenets of the radicals' creed. In the early

[14]Polubarinova-Kochina, "Zhizn'," p. 10.

[15]Prudnikov, "A. N. Strannoliubskii," p. 124.

[16]Much has been written recently on women's movements in Russia in the latter half of the nineteenth century. See, for example, Richard Stites' excellent book, *The Women's Liberation Movement in Russia*; Rochelle Lois Goldberg, *The Russian Women's Movement: 1859–1917* (Ph.D. dissertation, University of Rochester, 1976). For more detail on strictly educational movements, see Stanton, *The Woman Question*; Shashkov, *Sobranie sochinenii*, vol. I; M. Peskovskii, "Universitetskaia nauka dlia zhenshchin," *Russkaia mysl'*, Nos. 11 and 12 (1886); Sophie Satina, *Education of Women in Pre-Revolutionary Russia* (New York: n.p., 1966); E. Likhacheva, *Istoriia zhenskogo obrazovaniia*, vol. II; S. N. Valk, ed., *Sankt-Peterburgskie vysshie (Bestuzhevskie) kursy* (Leningrad: Leningradskii Universitet, 1973). Daniel Brower's book, *Training the Nihilists* (Ithaca, N.Y.: Cornell University Press, 1975), which at first glance seems promising, dismisses the women as not playing a significant enough part in the radical movement to be included in his study (p. 38).

sixties, women formed a large part of an amorphous group of unofficial auditors who wandered in and out of the lecture halls in St. Petersburg institutions of higher education. According to the popular essayist L. F. Panteleev, who as a student was exiled to Siberia for his part in the student uprisings of 1862, the professors were not particularly sympathetic to the women, but they were not hostile either. "They had no suspicion that this was the beginning of a serious movement, but rather saw the striving [of the women] as simply a new fashion."[17]

At the same time, the Medical-Surgical Academy opened its doors to women as medical students on a more or less official basis. Among the eager enrollees were Nadezhda Prokofevna Suslova, who was the daughter of a wealthy freed serf and sister of Dostoevskii's mistress Polina; and Maria Aleksandrovna Bokova-Sechenova, who was the fictitious wife of P. I. Bokov, a member of revolutionary circles and friend of Chernyshevskii, Pisarev, and others.[18] These women not only were champions of female education, but also were active in the broader political scene. Along with Natalia Ieronimovna Korsini-Utina, who audited law and constitutional history courses at the university, they were members of the first Land and Freedom organization. This was one of the early populist groups, devoted to discussing and propagandizing the nihilist views on social issues. Thus, they were in intimate contact with all aspects of the progressive movement of their day.[19]

The number of women enrolled semi-officially as auditors or merely attending lectures unofficially grew steadily throughout 1860 and 1861. The faculties of Moscow and Dorpat universities had voted against the idea of admitting females, but hopes were high that the universities of St. Petersburg, Kiev, and Kharkov would soon fully open their doors to women, since all three admitted them semi-officially already.[20] But these hopes

[17]Panteleev, *Vospominaniia*, p. 215.

[18]For biographical sketches of these and other women of the 1860s and 1870s see P. I. Ariian, *Pervyi zhenskii kalendar'* (St. Petersburg: various publishers, 1899–1912).

[19]Panteleev, *Vospominaniia*, pp. 215, 317, n. 2.

[20]S. Ashevskii, "Russkoe studenchestvo v epokhe 60-kh godov (1855–1863)," *Sovremennyi mir*, No. 7–8 (1907), p. 21, n. 1.

were soon dashed. Student uprisings started in March 1861 and continued in waves throughout the year. At first, the demonstrations were partly gestures of solidarity with rioting Polish students, and partly protests against the inadequacies of the emancipation proclamation. Eventually, they broadened to include demands to fire incompetent or unfair professors, to allow student meetings which had been forbidden, and to permit students to band together in eating, reading, and lodging cooperatives.[21] The uprisings were put down with force, many students were expelled from the university, arrested and/or exiled to Siberia, and an era of reaction against students began.[22]

For women, the most unfortunate result of this reaction was that the universities and other institutions of higher education closed their doors to all but officially enrolled students. This had the effect of excluding all women from such institutions.[23] To be sure, in 1863 a series of women's pedagogical courses was started by professors E. K. Brandt of the Medical Academy, F. F. Petrushevskii of the Artillery Academy, and Sofia's champion N. N. Tyrtov of the Naval Academy. But these courses raised a furor because of rumors that immoral and materialist propaganda was being given to the students under the guise of anatomy and zoology lectures. In 1865, fearing government repression of the whole program, the founders canceled most of their scientific lectures. This left Russian

---

[21]Ashevskii, "Russkoe studenchestvo," *Sovremennyi mir*, No. 9 (1907), p. 64ff.; see also Brower, Chapters I and II.

[22]A. M. Skabichevskii, *Literaturnye vospominaniia* (Moscow: Zemlia i Fabrika, 1928), p. 76ff. Skabichevskii relates how once, after the put-down of the student disturbances in 1862, he inadvertently neglected to remove his student uniform before venturing away from the university. A child pointed him out to her grandmother, saying, "Look, there's one left!", sounding "exactly as if she were talking about cockroaches!" (p. 150).

[23]The one exception to this ruling was Varvara Aleksandrovna Kashevarova-Rudneva, a young Jewish woman who left her home and husband in Vitebsk to study medicine with the goal of helping Moslem women of the Bashkir nationality. The Orenburg provincial government made a special petition to the Minister of Education for her to be allowed to continue her studies, which was granted. (Meijer, p. 23; Satina, p. 84.) Although Kashevarova came from Vitebsk, which was not far from Palibino, Kovalevskaia does not seem to have known her.

women with little possibility for higher education except private tutoring and study abroad.[24]

As was mentioned earlier, two basic tenets of the young radical movement of the 1860s were a belief in the power of education to cure many social ills, and a belief that women are equal to men and therefore deserve all the same rights. The nihilist men were ready and eager to help the nihilist women to gain freedom and enlightenment. Prince Kropotkin mentions that it was common to meet young men who on principle would not rise when a lady entered the room, yet would walk clear across the city, cold and tired, to give free lessons to any young woman who sincerely wanted to study.[25] Richard Stites remarks that Kropotkin's story is "firmly based in reality." Out of the writings of Chernyshevskii, N. K. Mikhailov, and other democratic publicists

> emerged a widely accepted position of the intelligentsia on women which included a full, positive, ungrudging affirmation of the equality of the sexes and a belief that the two sexes ought to march together in some common struggle (itself in the process of being defined). . . .
> The men developed a kind of egalitarian chivalry toward women that was helpful, supportive, and respectful rather than patronizing.[26]

For many women, the informal yet rigorous tutoring provided by the men satisfied their need for education. Then, too, they joined communes and cooperatives which included women at all stages of intellectual development. The communes "acted as a kind of educational force; those who were better educated influenced the others."[27] For a number of women, however, this unofficial tutoring was not enough.

---

[24]Vorontsova, pp. 73–4.

[25]Kropotkin, p. 300.

[26]Richard Stites, "Women and the Russian Intelligentsia," in *Women in Russia*, p. 48. See also Bogdanovich, *Liubov' liudei*; and Vera Broido, *Apostles Into Terrorists* (New York: Viking, 1977).

[27]Sheila Rowbotham, *Women, Resistance and Revolution* (London: Penguin, 1972), p. 124.

They desired to prove their abilities to more than just their tutors and a circle of supportive friends. They wanted to matriculate officially, obtain legitimate degrees, and begin to help the common people in an active capacity. Consequently, they began to look more and more into the possibility of study abroad.[28]

Rumor had it that the Swiss universities would admit foreign women without entrance examinations or diplomas (an important point, since many Russian women had had no formal schooling). Starting in 1864, Russian women wanting advanced study began to consider Zurich as their mecca. First came several of those students who had started their education before the Russian universities were closed to them in 1862–63. Nadezhda Suslova was among them. She enrolled as an auditor in 1865, in 1867 became the first official female student, passed her exams in August of that year, and on December 14, 1867, became the first woman to receive a medical degree from a European university.[29] (Elizabeth Blackwell, an Englishwoman, had received her diploma from a Rome, New York medical school in 1849.)

Suslova returned in triumph with her new degree, and was welcomed with pride by her old friends—Maria Bokova-Sechenova, Natalia Korsini, and Maria Bogdanova (who had attended science lectures at the university and later at the pedagogical courses). Suslova was also greeted enthusiastically by young women who had heard of her accomplishment and were eager to ask her advice and encouragement. Among them were the Korvin-Krukovskaia sisters.

At first, it might seem strange that Anna and Sofia—well brought-up women of noble family and affluent circum-

[28]As early as October 1863 Nadezhda Suslova asked her sister Polina, who was abroad with Dostoevskii, to inquire if there was any possibility of a woman studying medicine in Paris. At the time, as Polina discovered, there was no such possibility. (A. P. Suslova, *Gody blizosti s Dostoevskim* [Moscow: Izd. Sabashnikovykh, 1928], pp. 65–6.)

[29]Meijer, p. 25. This book is a rich source of information on the nihilist and emancipationist women of the 1860s and early 1870s. Meijer deals with the Zurich Russian colony as a whole, but he (unlike Brower, for example) recognizes that the women played a vital role in the political debates of Russians at home and abroad, and so he focuses considerable attention on them.

stances—would readily find circles of nihilists into which to insinuate themselves. But it was not difficult. Nihilism had made its way into some of the most elegant drawing rooms in St. Petersburg, and it was not hard to identify fellow believers, even in a crowded room of conservative older people. As Kovalevskaia later recalled to Anna Carlotta Leffler:

> Oh, what a happy time it was! . . . We were so enthusiastic about the new ideas; so sure that the present social state could not continue long. We pictured to ourselves the glorious period of liberty and universal enlightenment of which we dreamed, and in which we firmly believed. Besides this, we had the sense of true union and cooperation. When three or four of us met in a drawing room among older people,—where we had no right to advance our opinions,—a tone, a glance, even a sigh, were sufficient to show one another that we were one in thought and sympathy . . . ready for self-sacrifice in the same cause![30]

The Korvin-Krukovskaia sisters soon became convinced that the "self-sacrifice" needed was the forfeit of their own personal happiness. They both decided that in order to study abroad at least one of them would have to forego the chance of a happy marriage and opt for a fictitious husband. Sofa seems to have been at least as determined on this point as Aniuta, but the pair decided that Aniuta would be the more obvious bride. It was unlikely that their father would consent to eighteen-year-old Sofa's marriage when twenty-five-year-old Aniuta was still unwed.[31]

That winter in St. Petersburg, the persistence and singlemindedness that characterized Sofia throughout her life were more apparent than ever. She rapidly immersed herself in plans to liberate not only herself and Aniuta, but also her cousin Zhanna Evreinova, and even Zhanna's Moscow cousin Iulia Lermontova. Zhanna wished to become a lawyer, but her parents were adamantly opposed to the idea of a woman looking for education. Outwardly, the unhappy Zhanna submitted; but she did not give up hope. She secretly persuaded her German

[30]Leffler, "Sonya Kovalevsky," pp. 160–61.

[31]Shtraikh, *Sestry*, p. 53.

tutor to teach her Latin and Greek, and clandestinely studied economics to prepare herself for eventual matriculation.[32]

Iulia's parents, unlike Zhanna's, were sympathetic to her desire for education. They had always obtained tutors and books for her, and did not object when she made the rounds of Moscow scientific institutions, without success, asking if any of them would allow her and another two cousins to study chemistry and medicine. The Lermontov parents drew the line, however, at permitting Iulia to go abroad alone, so she was determined to circumvent their authority if she could.[33]

Sofa, Aniuta, and Zhanna began to look around for possible fictitious husbands. They communicated their plans to Iulia, who lived chiefly in Moscow, so she was willing to be patient and await the outcome. The women believed that one or at most two fictitious husbands would suffice for all four of them. Single young women sometimes lived with their married sisters or cousins; anxious parents agreed to the arrangement in the hope that their spinster-offspring would find husbands.[34] Ideally, the ones to be married should be the oldest, since they would be fulfilling the role of chaperone. Thus, the four decided that twenty-five-year-old Aniuta and twenty-four-year-old Zhanna were more suitable than twenty-two-year-old Iulia and the baby of the group, eighteen-year-old Sofa.

But, as Aniuta and Zhanna discovered, possible fictitious husbands were not numerous. It was mentioned earlier that not all nihilist men approved of the idea of fictitious marriage. Many looked forward to loving, equal relationships with women, and had no wish to jeopardize their chance for personal happiness by tying themselves to unknown females, however worthy of liberation they might be. To be sure, there were one or two young men willing to enter into a fictitious

[32]Vorontsova, pp. 76–7. It was considered proper for Zhanna to want to learn a modern foreign language, since all well-born ladies knew at least one or two. But the classical languages were considered too scholarly for a young girl. Her parents would never have approved; hence her secrecy. For further details on Zhanna's life see V. Lestvitsyn, "Ot"ezd A. M. Evreinovoi," *Iaroslavskie gubernskie vedomosti*, 18 September 1880, p. 5.

[33]Lermontova, p. 375. For a biography of Iulia, see Iu. S. Musabekov, *Iulia Vsevolodovna Lermontova* (Moscow: Nauka, 1967).

[34]See, for example, Kovalevskaia's *Nigilistka*, VP 1974, p. 133.

marriage. But, to Aniuta's and Zhanna's chagrin, these men were eager to help Sofa rather than the older women.

Sofia's tutor Strannoliubskii had been so impressed with his pupil's mathematical and political level that he mentioned her to all his acquaintances.[35] He praised her intellectual ability, remarking upon her determination to study. Her fame spread, and one nihilist circle decided that one of their number should be the instrument of Sofia's liberation. Apparently, they thought only of Sofia rather than of Aniuta because of Sofia's more definite abilities and scientific interests.

According to the writer L. F. Panteleev, the circle decided on Panteleev's friend Ivan Grigorevich Rozhdestvenskii, an upright, progressive young man whose only drawbacks were extreme ugliness and a slight overfondness for beer. Rozhdestvenskii sought Sofia's consent to the proposed fictitious marriage. Sofia and Aniuta must have been fairly sure Rozhdestvenskii's attempt to win the agreement of their father would be unsuccessful. Rozhdestvenskii's financial prospects were dim, since he had quarreled with his wealthy father over politics, and his appearance was unprepossessing. The sisters had no alternative candidates, however, so they agreed that he should try.

Rozhdestvenskii's friends set about preparing him to ask Sofia's father's permission. The usual nihilist wardrobe being too informal for such an occasion, they bought him a used morning coat and top hat at the peasant market, and set him to practicing bows in front of a mirror. General Korvin-Krukovskii could hardly contain his astonishment when the ugly, ill-dressed Rozhdestvenskii came to his study and requested Sofa's hand. The general asked the young man what he did. Rozhdestvenskii answered with aplomb that he was a "free educationist," in plain words, a free-lance tutor. General Korvin-Krukovskii thanked him politely, but said that Sofa was far too young. He managed to contain his amusement until Rozhdestvenskii was almost out the door.[36]

[35]Vladimir Stasov, *Nadezhda Vasilevna Stasova* (St. Petersburg: Merkushev, 1899), pp. 177, 240; E. El' [pseudonym of E. Litvinova], "Iz vremen moego studenchestva," *Zhenskoe delo*, No. 4 (1899), p. 34.

[36]L. F. Panteleev, "Iz zhizni S. V. Kovalevskoi," *Birzhevye vedomosti*, 29 January 1916, p. 3; P. Ia. [Polubarinova] Kochina, *Sofia Vasilevna Kovalevskaia*

Rozhdestvenskii returned to his friends and reported the failure of the plan for Sofia's liberation. News spread, and reached the wider circle of Rozhdestvenskii's acquaintances. Possibly, it was in this way that the young nihilist Vladimir Onufrievich Kovalevskii (1842–1883) heard of the Korvin-Krukovskaia sisters' plight.

At the time, Kovalevskii was a publisher of scientific and political works, and a dilettante scientist, who had not yet done the work which would later make him famous in scientific circles.[37] Kovalevskii was the second son of a Russianized Polish landowner in Vitebsk and his Russian wife.[38] He had been educated in an exclusive boarding school where he learned to speak French, English, and German fluently. At the insistence of his father, he then went to a St. Petersburg institute for the study of jurisprudence, in order to prepare himself for government service. There he met such future nihilists as N. V. Shelgunov, M. L. Mikhailov, P. I. Iakobi and others, and decided that a career in the civil service did not appeal to him. He traveled abroad in 1861, met with Russian political émigrés, lived with Herzen's family in London for a year and tutored their daughter Olga.

Kovalevskii took some small part in the Polish rebellion in 1863, returned to Russia clandestinely, and fell into revolutionary circles there.[39] He began to participate actively in a circle

---

(Moscow: Nauka, 1981), p. 36, n. 1. It is interesting to note that Panteleev does not mention Aniuta at all here.

[37]See Shtraikh, *Sem'ia*; A. A. Borisiak, *V. O. Kovalevskii* (Leningrad: AN SSSR, 1928); L. Sh. Davitashvili, *V. O. Kovalevskii* (Leningrad: AN SSSR, 1946); Semen Reznik, *Vladimir Kovalevskii* (Moscow: Molodaia Gvardiia, 1978).

[38]Kovalevskii's older brother Aleksander (1840–1901) was an eminent embryologist about whom it was once said: "Ah, that Kovalevskii, he's a sharp fellow. He knows his [microscopic sea creatures] and [the history of revolutions], but he has not the slightest interest in anything in between." (K. A. Timiriazev, *Nauka i demokratiia*, p. 94.)

[39]There is some question as to how large a part he played. In her "Obshchestvennaia i literaturnaia deiatel'nost' S. V. Kovalevskoi" in *Pamiati*, p. 75ff., M. V. Nechkina attacks L. Sh. Davitashvili for consistently downplaying Vladimir's revolutionary activities. (In later editions of his biography Davitashvili corrected what Nechkina called his "blunders" with regard to Vladimir's politics.) In any case, Kovalevskii certainly traveled to Lvov during the height of the fighting, rescued his friend P. I. Iakobi, who had been

which included Maria Mikhaelis, the later-famous populist Petr Lavrov, the nihilist critic Varfolomei Zaitsev, and Zaitsev's nihilist sister Varvara. The circle's primary aim was to give material and educational aid to poor working women, and to set up women's employment cooperatives. The legal experience of Kovalevskii, who at one time had thought of becoming a lawyer, was helpful in drafting charters for the organizations. These would not be approved by the tsarist government unless they appeared innocent of any ties to nihilist circles.[40]

In 1864 Kovalevskii became engaged to Maria Mikhaelis, who was part of an illustrious family of intellectuals and revolutionaries. Maria's mother, Evgenia Egorovna Mikhaelis, was a writer and champion of women's rights. Maria's sister Liudmila was married to the revolutionary N. V. Shelgunov and shared all his activities, and her brother Evgenii had been arrested and exiled to Siberia as a result of the student demonstrations in 1862. She herself had been arrested and sentenced to a year's exile at the family estate because she threw a bouquet of flowers to Chernyshevskii when he was taken to Siberia in May 1864.

The marriage, however, never took place. On the morning of the designated day in September 1865, the two had a talk and decided not to go through with the ceremony. Maria's sister Liudmila remarked in her memoirs: "The wedding was broken off in the strangest manner. To their deaths neither one ever explained why they parted."[41]

This reticence on Maria's and Vladimir's part soon gave rise to speculation, and had unpleasant consequences for Kovalevskii. In April 1866, after Dmitri Karakozov took a shot at Tsar Alexander II, there was a roundup of politically suspect people in St. Petersburg and other large cities. In his report of the

---

wounded by tsarist troops, and hid him until the danger was past. (Shtraikh, *Sem'ia*, p. 42.)

[40]Shtraikh, *Sem'ia*, p. 44ff. For more information on the projects of the circle, which included the "Society for the Encouragement of Female Labor," see "Iz dnevnikov E. A. Shtakenshneider," *Golos minuvshego*, No. 4 (1916), p. 50ff. For a look at these projects from another point of view—that of the feminists—see Stites, *The Women's Liberation Movement in Russia*, p. 69ff.

[41]Shelgunova, *Vospominaniia*, vol. II, p. 119.

arrests in the widely-read émigré journal *Kolokol*, Herzen published Vladimir's name on a list of arrestees. As it happened, Herzen's information was wrong. Although it is now known that Kovalevskii was being watched by the tsarist secret police, he had never been seized. But in St. Petersburg, the discovery that Kovalevskii had not been kept in detention led to the suspicion that he had been released because he was a spy who had been arrested by mistake. Rumors circulated widely, and rival camps sprang up. Bakunin scoffed at the very idea, Herzen as usual vacillated, but the Russian émigré grapevine spread the rumor as if it were fact. When Bakunin challenged the accusers, one of the main pieces of evidence they used was the breakup of the Kovalevskii-Mikhaelis engagement: they said that Maria must have had some idea of her fiancé's supposed duplicity.[42]

Kovalevskii's biographers leave no doubt that the accusation of treachery was false. Nevertheless, they feel that the charge makes the question of why the wedding was called off assume more importance than it would ordinarily have. Shtraikh speculates that Maria discovered that Vladimir "was possessed neither of a strong character nor strong political convictions."[43] Davitashvili, on the other hand, hints that she was repulsed by Vladimir's deep-seated neuroticism, "his peculiar nature."[44]

Since Kovalevskii remained friends with the Mikhaelis family all his life, and since his later behavior with Sofia was often unbalanced, Davitashvili's conjecture is probably closer to the truth. Vladimir was a kind, honest, politically progressive young man with a "truly brotherly way of relating to women."[45] But as will be seen in later chapters, he had an unstable and high-strung personality.

[42]S. Ia. Shtraikh, "A. I. Gertsen i V. O. Kovalevskii," *Istoricheskie zapiski*, 54 (1955), pp. 448–63. Even at the time, more rational people realized that the charge was false. See, for example, Timiriazev, p. 94; P. D. Boborykin, *Vospominaniia* (Moscow: Khudozhestvennaia Literatura, 1965), vol. I, p. 364. The rumors were persistent, however, and cropped up throughout Kovalevskii's life.

[43]Shtraikh, *Sem'ia*, p. 49.

[44]Davitashvili, p. 31.

[45]Litvinova, p. 31.

Vladimir Kovalevskii made the acquaintance of the Korvin-Krukovskaia sisters through Maria Bokova. The women soon discovered that he supported the idea of education for women, and was willing to be used for that purpose. Kovalevskii's hand had been solicited in fictitious marriage once before, in September 1865, by the nihilist and emancipationist Varvara Aleksandrovna Zaitseva. Unfortunately for her, she chose to accost him only a few days after his engagement with Maria Mikhaelis was broken off, and he refused her unconditionally.[46]

This time, however, Kovalevskii was receptive to the idea of a fictitious marriage. He had heard much about Sofia from his friends, who included Bokova, her fictitious husband Bokov, her lover Sechenov, and Sofia's rejected suitor Rozhdestvenskii. Aniuta does not seem to have been as well-known in radical circles as her younger sister, so it was perhaps not surprising that Vladimir, himself a dilettante scientist, announced that he preferred to marry Sofia.[47]

Kovalevskii's biographers tend to pity him, claiming that he fell in love with Sofia from the beginning, yet had to hide that love for fear of angering or offending her. Davitashvili goes so far as to say: "It was a fictitious marriage, but the love and devotion on V. O.'s side was greater than in the majority of totally normal marriages." A. A. Borisiak is more moderate in his evaluation of Kovalevskii's emotions before the wedding. He also feels that Vladimir was in love from the beginning, but he finds the source of that sentiment in Sofia's "unusual love for science. Because he saw this in her, he wanted to help her particularly rather than her sister."[48]

Kovalevskaia's friend and biographer, Elizaveta Litvinova, discounts the suggestion that Vladimir was infatuated with Sofia from the beginning. She feels that Kovalevskii was

[46]Shtraikh, *Sem'ia*, pp. 51–2, 77–8. Zaitseva never forgot what she considered to be an insult, and turned her future husband P. I. Iakobi and her brother, the revolutionary publicist V. A. Zaitsev, against Kovalevskii. Both had formerly been good friends of Vladimir.

[47]Leffler, "Sonya Kovalevsky," pp. 165–66. There does not seem to be solid evidence for the story, often repeated in the biographical literature, that Vladimir first accepted Aniuta, and then, having met Sofia, declared that he would have her or no one.

[48]Shtraikh, *Sem'ia*, pp. 136–67; Davitashvili, pp. 36, 41; Borisiak, p. 12.

motivated by his intellect rather than his heart: "Kovalevskii, as an intelligent person, noticed the young girl's prominent talent, and wanted to open the road to science specifically to her."[49]

The explanation that Kovalevskii wanted to marry Sofia to ensure her scientific career is not as farfetched as it at first seems. Recall that Aniuta had assured the professor the women had initially approached that it did not matter which one of them he married, because his future wife could chaperone the others. But in fact, this was not necessarily the case. Fathers could refuse to allow daughters to accompany married sisters. Some did refuse, often at the last minute. Kovalevskii might not have shared the Korvin-Krukovskaia sisters' optimism on this question. Perhaps he wanted to marry Sofia to make sure that she at least would be able to go abroad.

Although Kovalevskii was willing to marry Sofa, there is some evidence that he was not at first convinced of the necessity of the marriage. Maria Mikhaelis' sister Liudmila Shelgunova recalls that Vladimir sent a letter to her mother, Evgenia Mikhaelis, saying "that he had met an amazing girl Korvin-Krukovskaia, who wanted to study, but her parents would not let her, and he asked my mother to take her and give her refuge. My mother responded to this letter with full sympathy."[50]

Before arrangements with Evgenia Mikhaelis could be finalized, however, Vladimir and Sofia decided to get married. For Sofia, the fictitious marriage plan had an obvious advantage over Vladimir's alternate proposal. If all went well, the wedding presented at least the possibility of liberating Aniuta, Zhanna, and Iulia as well as Sofia. Taking refuge with the Mikhaelis family would have meant abandoning the other women. Sofa felt a sense of comradeship with them, so she was reluctant to agree to a plan which had no provision for her fellows.

Neither Sofia nor Vladimir seems to have had pressing reasons for entering into their marriage. Kovalevskaia's friend

[49]Litvinova, p. 31.

[50]Shelgunova, *Vospominaniia*, vol. II, p. 119.

Elizaveta Litvinova considered the whole idea of a fictitious marriage to be foolish, and looking back, felt that in Sofia's case it was absurd. Litvinova believed that General Korvin-Krukovskii could have been persuaded to allow Sofia to study abroad. In her opinion, Sofia succumbed to the marriage idea because it was exciting, and fashionable in certain radical circles.[51]

It is possible that Litvinova was right, and that Sofia eventually could have cajoled her father into letting her go to Germany or Switzerland. In the first place, she was her father's favorite, and he therefore found her generally more persuasive than Aniuta. Moreover, Sofia had the advantage that her plans were more definite than had been Aniuta's when she unsuccessfully urged her father to let her study in St. Petersburg.

Sofia had tested abilities in the sciences and mathematics. Her tutor Strannoliubskii and her early champion Professor Tyrtov would have supported her request for formal university training, and it is probable that Vasilii Vasilevich would have given in. Aniuta and Iulia could have accompanied her, leaving only Zhanna to worry about. In light of the tragedy which Sofia's fictitious marriage turned into, it is unfortunate that the women did not consider this alternative.

Sofia seems never actually to have requested her father's permission to study abroad. Instead, she set to work persuading Vasilii Vasilevich to agree to the match. As expected, he was reluctant to give his consent, although he probably would have agreed to a marriage between Kovalevskii and Aniuta. As Litvinova waspishly put it: "Anna and her girlfriend [Zhanna] had already succeeded in completely irritating their parents and so fatiguing them with their eccentricities that their parents would have even considered not particularly good matches as a happy release."[52] Moreover, Aniuta was already twenty-five, and close to being an "old maid." Sofa, on the other hand, was young and attractive, and the general thought she would be able to make a better marriage than to an impecunious book publisher.

[51]Litvinova, p. 30.

[52]*Ibid.*, pp. 31–2.

General Korvin-Krukovskii equivocated, and then asked Vladimir to wait until Sofa was older. Unused to being thwarted, the headstrong Sofa took matters into her own hands. She waited until the Korvin-Krukovskii family was sitting down to dinner with a number of illustrious guests— Uncle Briullov (a prominent diplomat), two well-known professors, and three generals—and then slipped out of the house. A note was delivered into her father's hands, saying she was at Vladimir's lodging without a chaperone, and would not come home until the general consented to the wedding. Vasilii Vasilevich was outfaced. Sofia was in effect threatening to compromise herself publicly. Her father had no choice but to send for Sofia and her "fiancé," and officially present Kovalevskii to the assembled guests.[53]

In the months before the wedding in September 1868, Sofia studied diligently, and inspired Vladimir with the desire to work as well. He had always been inclined toward the sciences, but he was unambitious and tended to seize on every excuse not to settle down to serious study. He occupied himself with his press, and told himself that he could turn to scientific work later. Kovalevskii's publishing venture was acclaimed as a public service by the Petersburg intelligentsia. Once, he had even gone so far as to reprint Herzen's banned *Kto vinovat?* (Who is to Blame?) by the simple expedient of leaving off the author's name. In spite of the press's popularity among the "people of the sixties," however, the business was unprofitable and needed constant attention. Sofia, Aniuta, Maria Bokova-Sechenova, and other friends and relations did some of Kovalevskii's translations for him, often without pay. But the business continued to languish.[54]

During the engagement, Sofa took her fiancé in hand. She taught him mathematics, and they studied physics together. Kovalevskii later told Shelgunova that he only studied because "he was ashamed in front of his wife by his ignorance."[55] Cer-

[53]Leffler, "Sonya Kovalevsky," pp. 168–70.

[54]Shtraikh, *Sem'ia*, p. 74; N. A. Serno-Solovievich, 3 November 1865 letter to V. V. Ivasheva (Cherkesova), in M. K. Lemke, *Ocherki osvoboditel'nogo dvizheniia* (St. Petersburg: O. N. Popovaia, 1908), p. 225.

[55]Shelgunova, *Vospominaniia*, vol. II, p. 119.

tainly, he was in awe of his diminutive fiancée. To his brother Aleksander, he wrote: "With every passing day I have more opportunity to be amazed at the capabilities that can fit into such a young head." Later in the same letter, he voiced his hopes that Sofa would be the making of him: "I think that this meeting will make of me a decent person, that I will give up publishing and start to study, although I cannot conceal from myself that hers is a nature a thousand times better, smarter, and more talented than mine."[56]

Indeed, by now Kovalevskii was infatuated, not only with his bride-to-be, but with her whole circle as well. He appears to have been fascinated by and drawn to the dynamic new force of womankind which Sofa, Aniuta, Maria Bokova-Sechenova, and Nadezhda Suslova represented. After the sisters left Petersburg in late April 1868, he wrote to Palibino, addressing them both:

> Truly, acquaintance with you makes me believe in *Wahlver-wandtschaft* [the meeting of minds or souls], so swiftly, immediately and genuinely did we come together and, on my side at least, become friends. . . . Truly, judging in the most objective manner, without childish enthusiasm, one can say almost certainly that Sofia Vasilevna will be a wonderful doctor or scholar in some area of the natural sciences; further it is very probable that Anna Vasilevna [Aniuta] will be a talented writer, that Nadezhda Prokofevna [Suslova] and Maria Aleksandrovna [Bokova] will be excellent doctors, that Iv[an] M[ikhailovich] Sechenov will always remain (for some) or become (for others) our mutual and best friend. . . .[57]
>
> Thus you see that I merge with your interests not only mine, but the interests of the above-mentioned personalities as well . . .

[56]May 1868 letter, reprinted in Shtraikh, *Sestry*, pp. 66–7.

[57]I. M. Sechenov (1829–1905) was a nihilist, champion of women's educational rights, and eminent physiologist. He was Maria Bokova's common-law husband for many years, and later her legal husband. He and P. I. Bokov (Maria's legal husband) shared an apartment with her, entertained guests together, and even sent out joint invitations and announcements when Maria successfully completed her medical examinations. (Bogdanovich, *Liubov' liudei*, pp. 59, 426.) At first the Korvin-Krukovskaia sisters seized upon Sechenov as a possible fictitious husband for Aniuta, but he and Maria acted obtuse whenever the subject was hinted at. This behavior irritated the Korvin-Krukovskaia sisters no end.

Therefore you should look on me not as a man who has done you a favor, but as a comrade who is striving toward one goal together with you. That is, I am necessary to you just as much as you are necessary to me. It follows that you should slight me and use me *en consequence* in everything you can think of, without worrying about inconveniencing me—I work here [in St. Petersburg] as much for you as for myself.[58]

It is clear from this passage that Vladimir's concept of fictitious marriage differed sharply from that of the Korvin-Krukovskaia sisters. He seems to have anticipated a union much like that of his friends Petr and Maria Bokov. The Bokov marriage had been proposed as the means of freeing Maria from her tyrannical father, but Petr had never had the intention of abandoning his fictitious bride as soon as the knot was tied. The pair lived together as friends, Petr helped Maria with her studies, and eventually the marriage was consummated.

Aniuta and Sofia, on the other hand, had in mind a more superficial concept of fictitious marriage. They expected Vladimir to fulfill his role as bridegroom, conduct the sisters to Germany or Switzerland, and then disappear from their lives. Aniuta especially seems to have been nonplussed when Kovalevskii entered so enthusiastically into their lives and concerns. Hence, in her letters she repeatedly protested that Vladimir was putting himself out for her too much.[59] This dif-

---

[58]Quoted in Shtraikh, *Sestry*, pp. 56–7. Stillman mocks this letter, sarcastically deriding the "convinced materialist" for his talk of kindred souls, and asserting that "Kovalevsky had violated a canon of radical ethics" by displaying his love for Sofa so blatantly and insisting on marrying her. (Introduction to Kovalevskaya, *Russian Childhood*, p. 9.) But Stillman quotes the letter from its abridged version in *VIP 1951* (pp. 485-87), and consequently misses the point. This portion of Kovalevskii's letter is addressed to both Aniuta and Sofa (a passage on Aniuta's writing precedes it). Thus, it speaks more of his previous loneliness and present happiness in new-found friends than it does of any depth of personal feeling for Sofia. A letter written to his brother Aleksander in May would seem to confirm this: "I met two girls and then their family almost accidentally this winter—they are from our own Vitebsk. . . . I made friends very quickly with the daughters [of the family] and found them to be very unusual personalities. . . . Don't think that these are simply girls with good intentions or wishes . . . they are hard-working and astoundingly advanced beings." (Quoted in Shtraikh, *Sestry*, pp. 63–4.)

[59]Kovalevskii was answering such protestations in the last passage of the letter quoted above.

ference in interpretation of the meaning of fictitious marriage would cause many problems for Vladimir and Sofia in the years to come.

Plans for the wedding went forward, although the atmosphere at Palibino was strained. "Father and I see each other only at lunch and dinner," Sofia wrote to her fiancé in July, "and these short meetings lead only to the exchange of caustic remarks; I, by the way, am more silent."[60] Sofia was content to have these cold relations with Vasilii Vasilevich. Her one fear, as she wrote to Kovalevskii, was that her father would relent and show tenderness to her, in which case she was not sure of her ability to resist his appeals to her to postpone the wedding.[61]

Unfortunately for Sofa, in light of later events, General Korvin-Krukovskii did not relent. He was too upset by his daughter's actions to be able to control himself, and so did not speak to Sofa. Early autumn saw Palibino in a welter of preparations for the wedding. Aunt Briullova sent Sofia's German cousin Sophie Adelung a running commentary, from which it seems that Sofa was happy. She spent almost all her time with Aniuta, Zhanna (who had come to Palibino for the celebration), and Vladimir, virtually ignoring all the relatives and friends who had gathered at the estate.

Aunt Briullova liked Kovalevskii, who impressed her "as a man one can trust." Even General Korvin-Krukovskii was finally induced to be cordial to him, although the general, wrote Aunt Briullova:

> could not forgive him for taking away his favorite daughter. Poor V. V. is so depressed that I cannot look at him without becoming depressed myself. Naturally, he is very worried about Sofa's future, and thinks that she is still too young to feel a genuine love and will soon snap out of her intoxication. Lisa [Sofa's mother Elizaveta Fedorovna] is much more hopeful, and Sofa is the picture of happiness.[62]

[60]*VIP 1951*, p. 206.

[61]*Ibid.*, p. 210.

[62]18 September and 22 September 1868 letters in Adelung, pp. 414–15. It

Probably, Sofa's true state of mind could have been described more properly as triumphant, or even gleeful. She had acquired a fictitious bridegroom, would be living in her own apartment in St. Petersburg, and would soon be going abroad to study. Her only worry was Aniuta's inability to escape from Palibino with her.

Vasilii Vasilevich was upset about Sofa's choice of husband and her deceitful way of gaining her father's consent to the match. Consequently, he resisted Sofa's cajolery, and refused to let Aniuta go to St. Petersburg with the newlyweds after the ceremony on the twenty-seventh of September. The most Sofa could obtain was a promise that Aniuta would be allowed to come to her later in the fall. She assured Aniuta that she would try to find a suitable fictitious husband for her in the meantime.

Thus, in late September 1868, little Sofa Korvin-Krukovskaia became Sofia Kovalevskaia. At the age of eighteen, she left her childhood behind her, and was ready to embark upon a new life. It would prove to be difficult to obtain higher education, but Sofia was used to overcoming obstacles. She had persuaded her father to let her study calculus in spite of his previous opposition. Now, she was ready for the challenge of university-level studies.

---

should be noted that only Aniuta, Zhanna, and Vladimir knew the nature of Sofia's and Vladimir's marriage.

four

IN SEARCH OF
EDUCATION

As soon as Vladimir and Sofia arrived in St. Petersburg after their wedding, they plunged into a life of hard work interspersed with pleasant social encounters with other nihilists.[1] Sofia resumed mathematical lessons with Strannoliubskii, and set about finding teachers for herself in other subjects as well. She planned to study physics and the natural sciences, and hoped to audit courses at the Medical-Surgical Academy clandestinely.

The day after the newlyweds arrived in the capital, Kovalevskaia was conducted to Sechenov's physiology lecture by an escort consisting of her husband, Petr Ivanovich Bokov (Maria's fictitious husband), and her uncle Petr Vasilevich Korvin-Krukovskii. The men accompanied her to shield her from the attention of the university authorities, and because they feared unpleasant comments from men in the class. Most of the other medical students were true men of the sixties, however, and unobtrusively moved to hide Sofia from the eye of any government inspector who should happen to glance into the classroom. Sofia was delighted with Sechenov's lecture, and wrote to Aniuta that with her new studies, "my real life begins."[2]

Relations between Vladimir and Sofia at this stage of their marriage were idyllic, though of course at this time there was no sexual component in their relationship. As Stillman de-

---

[1]Sofia and Vladimir set up a five-room apartment on Sergeevskaia Street.

[2]29 September 1868 letter in *VIP 1951*, p. 223.

scribes it, their life together "resembled a pre-adolescent's fantasy. He lavished on her an inexhaustible fund of tenderness, pride and solicitude . . . and she was inseparable from him and addressed him constantly as 'brother.' "[3] In fact, they lived much like Vera Rozalskaia and Dmitri Lopukhov in Nikolai Chernyshevskii's famous novel *What is to be Done?*. In the context of the 1860s, the relationship was not odd at all.

There were several such marriages among the early nihilists—marriages in which the husband made a conscious decision to defer to his wife's needs and help her attain full development of her personality and intellect. Some of the marriages were fictitious. Others began that way and then, like those of the Kovalevskiis and the Bokovs, were consummated later. Still other marriages were open arrangements, where the wife had a right to take lovers. The marriages of the revolutionary publicist N. V. Shelgunov and that of Chernyshevskii himself were of this last type.[4]

It is possible that Sofia and Vladimir consciously modeled their behavior on that of the characters of *What is to be Done?*. On the other hand, they might have taken their cues from the couple whose marriage was probably Chernyshevskii's model: Maria and Petr Bokov. In any case, Sofia and Vladimir followed the pattern for fictitious marriage outlined in *What is to be Done?* and practiced by people of the sixties. They were tender and affectionate to one another in public and, judging by Kovalevskaia's letters to Aniuta, in private as well.[5] But they had separate rooms and would not think of intruding on each other's seclusion without permission. Moreover, as in other

---

[3]Stillman, Introduction to Kovalevskaya, *Russian Childhood*, p. 10. Elsewhere, Stillman seems to imply that this form of address was peculiar to and indicative of abnormality in Sofia's relationship with Vladimir. But, as Ekaterina Kuskova notes in her commentary on Kovalevskaia's youthful letters, "in the sixties and seventies people of similar ideas frequently considered each other brothers in spirit, and so called each other brothers and sisters." ("Pis'ma S. V. Kovalevskoi 1868 g.," *Golos minuvshego*, No. 2 (1916), p. 227, n. 2.)

[4]See Bogdanovich for entertaining accounts of famous fictitious marriages of the sixties.

[5]"Pis'ma S. V. Kovalevskoi 1868 g.," *Golos minuvshego*, No. 3, 4 (1916), *passim*.

relationships of this type, Vladimir deferred to Sofia, and put her interests before his own. Along with other idealistic men of the sixties, he felt that since women had been oppressed for so long, a special effort was needed: "at first men should accept second place, to give women time to outgrow their past servitude."[6]

Sofia was stimulated by her life in St. Petersburg both scientifically and politically. She arranged to study chemistry with N. N. Zinin, and physiology and anatomy with Sechenov. She tried to get F. F. Petrushevskii, who had once been involved in the women's pedagogical courses, to teach her physics, but he had withdrawn from anything connected with radical activities. In an attempt to persuade him, Vladimir Kovalevskii asked the physicist's assistant, P. P. Fan-der-Flit, to intercede. Fan-der-Flit's wife Polina was a cousin of the great revolutionary democrat N. G. Chernyshevskii. Polina introduced Sofia to Chernyshevskii's wife Olga, to Olga's two sons, and to her own brother, the radical Aleksander Nikolaevich Pypin.

Kovalevskaia was impressed by Olga Chernyshevskaia. She visited the family often, and remained friends with the Chernyshevskiis and the Pypins for the rest of her life. (She was in the process of writing a novel about the Chernyshevskiis when she died.) This friendship in itself was a political act, since even in exile Chernyshevskii was considered dangerous by the tsarist government.[7]

Although Kovalevskaia socialized with these and other people of the sixties in the evenings, her days were devoted to study. Strannoliubskii came several times a week. He often spent five hours at a time teaching higher mathematics to his talented pupil.

As a result of this intensive exposure to the discipline, Kovalevskaia decided that it was mathematics and only mathematics that she wished to study. She had given medicine a fair try, but her heart was not in the work. As she explained in a letter to Aniuta: "I am only happy when I am immersed in

[6]Broido, p. 25. See also Bogdanovich, p. 15.

[7]Vorontsova, pp. 96–7; Nechkina, p. 80.

contemplation; and if now, in my best years, I do not exclusively study my favorite subject, maybe I will never make up the lost time. I am convinced . . . that one lifetime will hardly be enough for all that I can do on my chosen path."[8]

Mingled with this excitement about mathematics and her nihilist friends, however, was uncertainty about her relationship with Vladimir Kovalevskii. Sofia was apparently finding it difficult to reconcile fictitious marriage in the style of *What is to be Done?* with the impersonal, formal arrangement she had envisioned. She could not rid herself of a vague feeling of dishonesty. Litvinova recounts that in those first months in St. Petersburg, Kovalevskaia would blush whenever anyone mentioned her marriage or called her by her married name.[9]

Sofia's letters to Aniuta are filled with references to her unhappiness. Usually, she writes that she is sad because she does not have Aniuta with her. But at the end of October, Sofia is more honest with herself:

> I am very unhappy, nothing is going right for me . . . It seems to me . . . that in my new relationship there has already arisen some kind of falsity—or not so much falsity as imperfect sincerity on my side, and on the other side also dissatisfaction.[10]

It would be a mistake to take Sofia's protestations of unhappiness at face value. Throughout her life, Kovalevskaia had a tendency to overdramatize herself and the situations in which she was involved. She would launch into some theatrical tirade or self-pitying monologue, and become carried away with her own eloquence. Worse still, she carried her friends with the force of her personality, so that they became convinced of the veracity of her flights of fancy.

Kovalevskaia's acquaintances remember occasions when she unconsciously duped them in this way. She would claim that she was deathly unhappy or ecstatically pleased about something. But just when she had worked herself and her friends up to a fever pitch of excitement, the essential rationality of her

[8]"Pis'ma S. V. Kovalevskoi," *Golos minuvshego*, No. 4 (1916), p. 88.

[9]Litvinova, pp. 32–3.

[10]"Pis'ma S. V. Kovalevskoi," *Golos minuvshego*, No. 4 (1916), p. 87.

nature would reassert itself. She would come back down to earth, and begin to act reasonably again. Moreover, she would view the credulity of her friends with surprise and even disdain. She could never understand why they took her dramatics so seriously.[11]

Aniuta, who possibly knew Sofia better than anyone else would ever know her, was not deceived by her younger sister's artistically written laments. Aniuta repeatedly chided her for exaggeration. But her remonstrances only seem to have goaded Sofia to further extravagancies of expression.

Kovalevskaia's letters to Aniuta during the first months of her marriage are curious documents. Invariably they begin with a long, sentimental protestation of undying love for her "dear darling," "dear priceless," "dear darling incomparable," "wonderful" Aniuta. These passages are almost feverishly unhappy in tone, and speak of Sofia's inability to exist apart from her "spiritual mother." But then, after the obligatory half-page of lamentation, Sofia's letters change tone completely. She writes excitedly of her plans, studies, new friends, and even her relationship with Vladimir Kovalevskii. She is joyful, and looking forward to each new experience.[12] In fact, Kovalevskaia makes it clear that, except for a slight uneasiness about Vladimir, her life suits her well. If only Aniuta, Zhanna, and Iulia could join her when she went abroad, everything would be fine.

Sofia turned her attention toward aiding her sister, cousin, and Iulia Lermontova to escape from their parents. Zhanna's case was the most hopeless. Her father had "declared he would rather see her in her grave than in a university."[13] As if that were not enough, he was encouraging the dishonorable attentions of the Grand Duke Nikolai Nikolaevich, who was attracted to Zhanna. As Kropotkin explains: "Such was the influence of the court upon St. Petersburg society that if one of

[11]There are examples of this behavior in the memoirs of Mendelson, Litvinova (and also her pseudonym "E. El' "), and Leffler; and in Ellen Key, "Sonja Kovalevska," *Drei frauenschicksale* (Berlin: Fischer, 1908), pp. 7–69.

[12]See "Pis'ma S. V. Kovalevskoi," *Golos minuvshego,* No. 3 and 4 (1916), *passim.*

[13]Stillman, Introduction to Kovalevskaya, *Russian Childhood,* p. 13.

the grand dukes cast his eyes upon a girl, her parents would do all in their power to make their child fall madly in love with the great personage, even though they knew full well that no marriage could result from it."[14]

Zhanna was desperate. She resolved to approach the revolutionary Jacobin, Petr Nikitich Tkachev, whom she knew, to propose a fictitious marriage to him. Sofia was opposed to the idea because she did not like Tkachev, "although of course I don't try to dissuade her, and won't say a word about it to brother [Kovalevskii]."[15] Zhanna did not care about Tkachev's personality, however. As she wrote to her cousin Iulia: "We are looking for men who, like us, are fervently dedicated to the cause, whose principles would be identical to ours, who would not marry us but liberate us."[16] Tkachev was not given the opportunity to liberate Zhanna—he was arrested before the couple could make concrete plans.[17]

Kovalevskaia tried to persuade the elder Evreinovs to let Zhanna go abroad under her chaperonage, but to no avail. She was more successful with Iulia Lermontova's parents who, after much hesitation, allowed her to follow the Kovalevskiis abroad.[18] Aniuta came with Vladimir and Sofia when they left St. Petersburg in April of 1869, and Iulia came in the fall.

Zhanna joined the group later, in a most spectacular manner. She was increasingly depressed by the attentions of the grand duke and her father's attempts to induce her to be complaisant. She even wrote to Sofia that she thought she would drown herself. But Vladimir put her in touch with V. Ia. Evdokimov, who was the manager of the bookstore where Kovalevskii sold most of his scientific and progressive books. Evdokimov, who was involved with the importation of illegal political literature into Russia, gave Zhanna money and the names of some people who could smuggle her over the Polish border. She fled on

[14]Kropotkin, p. 144.

[15]10 October 1868 letter to Aniuta, "Pis'ma S. V. Kovalevskoi," *Golos minuvshego*, No. 4 (1916), p. 82.

[16]Quoted in Knizhnik-Vetrov, *Russkie deiatel'nitsy*, p. 153.

[17]Shtraikh, *Sem'ia*, p. 146.

[18]Lermontova, p. 376.

foot at night, and eventually joined the Kovalevskiis, Aniuta, and Iulia in Heidelberg.

An unanticipated consequence of Zhanna's flight was the arrest of Evdokimov and the bookstore owner Cherkesov. It seems that Zhanna unwisely sent Evdokimov a letter of thanks which included an account of a workers' meeting in Berlin and animadversions on the tsarist government. The letter was opened by the censors, and led to several arrests.[19]

Sofia and Vladimir had different motives for leaving Russia. Sofia was looking forward to settling down to serious work. "I cannot imagine a happier existence than a quiet, modest life in some forgotten corner of Germany or Switzerland, among books and studies," she wrote to Lermontova.[20] Kovalevskii, on the other hand, left Russia to escape his creditors and the ruin of his publishing business. Yet under his wife's influence he too soon began to work seriously.[21]

First, Sofia and Vladimir traveled to Vienna, where Vladimir hoped to study geology and the fossil collections of museums. Sofia did not expect to be allowed to attend lectures in Vienna, but on principle she attempted to obtain admission to the university. To her surprise, she discovered that the physicist Professor Lange was willing to accept her as a student. She could not find any cooperative mathematicians, though. Since in the nihilist fashion Vladimir was putting Sofia's needs above his own, the pair moved on to Heidelberg to seek mathematicians for Sofia.[22]

Aniuta had accompanied the Kovalevskiis to Vienna, but her political aspirations soon supplanted her literary strivings. She decided to move on to Paris to contact the workers' movements there. Since Vasilii Vasilevich had only permitted Aniuta to go

---

[19]"Iz dnevnikov E. A. Shtakenshneider," *Golos minuvshego*, No. 4 (1916), p. 73, and note to that page by *Golos minuvshego* editor V. I. Semevskii; "V. O. i A. O. Kovalevskie," (letters) in *Nauchnoe nasledstvo* (Moscow: AN SSSR, 1948), vol. I, p. 226. Hereafter this collection of letters, many of which are from Vladimir to Aleksander Kovalevskii, is cited as *N N*.

[20]*VIP 1951*, p. 235.

[21]Shtraikh, *Sem'ia*, p. 149; Borisiak, p. 17.

[22]Letter to Lermontova, *VIP 1951*, p. 237.

abroad on the understanding that she would stay with Sofia, and since he was giving each of his daughters a thousand rubles a year for expenses, they decided not to make Aniuta's destination known to their parents. They made arrangements for Aniuta to send letters to Palibino through Sofia, who would mail them from Heidelberg. The sisters hoped that this would preserve the fiction that they were together.[23]

Life in Heidelberg those first few months was idyllic, both scientifically and personally. Kovalevskaia arrived in May 1869, and went immediately to the university from the train station.[24] At first, she encountered the usual excuses and bureaucratic runaround that university personnel at that time normally gave women—that is, if they did not refuse them outright. Part of the problem seemed to be that some unknown woman had told the university officials that Sofia was a widow, while Sofia maintained that she was married. The conflicting reports of Kovalevskaia's marital status made Heidelberg university administrators suspicious that she was some sort of adventuress, harboring evil intentions toward the virtue and pocketbooks of the academics of the town.

Fortunately, the combination of Vladimir's visit to the Rector and the illustrious name of Kovalevskii (his brother Aleksander was already famous for his embryological work) was enough to persuade the Heidelberg administrators to give provisional permission. Sofia would have to approach the professors individually, but if they agreed, she would be able to attend their lectures without hindrance.[25]

Heidelberg professors in the natural sciences and mathematics proved to be willing to accept Sofia. She attended eighteen

[23]*VIP 1951*, p. 521, n. 331(1); Leffler, "Sonya Kovalevsky," p. 184.

[24]Manuscript memoirs of Kovalevskaia's friend, the physicist Sergei Ivanovich Lamanskii, in fond 280 (Ekaterina Pavlovna Letkova-Sultanova), ed. khr. 340 of the Tsentralnyi Gosudarstvennyi Arkhiv Literatury i Iskusstva (hereafter called TsGALI), Moscow, USSR. Lamanskii attended some classes with Kovalevskaia in Heidelberg, and later was her business representative in Russia.

[25]Letter to Lermontova, 28 April 1869, pp. 237–38, and letter from Vladimir to his brother, 20 June 1869, p. 495, n. 238(1), both in *VIP 1951*; Lamanskii, TsGALI, f. 280, ed. khr. 340, l. 1.

to twenty-two hours a week of lectures during the three semesters she spent at Heidelberg. She took physics with the famous physical chemist Gustav Kirchhoff, and worked in his laboratory as well. She studied physiology with Hermann Helmholtz, and mathematics with the eminent professors Leo Königsberger and Paul DuBois-Reymond.[26]

In the fall of 1869, Iulia Lermontova arrived with the intention of studying chemistry. Sofia persuaded the woman-hating chemist Wilhelm Bunsen to let them both work in his laboratory. Bunsen had once said he would never allow a woman, above all a Russian woman, into his laboratory. Sofia cajoled him into relenting, and the cantankerous bachelor was furious when he grasped the fact that he actually had given two Russian women permission to invade his sanctum. "Now *that woman* has made me eat my own words," he later told Karl Weierstrass, Kovalevskaia's adviser in Berlin. Bunsen resented Sofia for, as he thought, making a fool of him, and later spread scandalous stories about her.[27]

Bunsen's jaundiced opinion was not shared by Sofia's other professors, who wrote enthusiastic letters to their colleagues at other universities. Kovalevskaia rapidly gained a reputation in the town of Heidelberg as well. In her memoirs, Lermontova recalls that people would sometimes stare at Sofia on the street, and that the newspapers featured articles about her.

In spite of this attention, Sofia was quiet and withdrawn, and rarely put herself forward. "This modest demeanor," Iulia noted:

> pleased her German professors very much, because in general they set great store on modesty in women, especially in such a prominent woman as Sofa, who in addition was studying such an abstract subject as mathematics.[28]

---

[26]Kovalevskaia's meticulous notes for these and most other lectures she attended during her student years are preserved in the archives of the Institut Mittag-Leffler of the Royal Swedish Academy of Sciences, Djursholm, Sweden; hereafter referred to as IML.

[27]G. Mittag-Leffler, "Weierstrass et Sonja Kowalewsky," *Acta Mathematica*, 39 (1923), p. 135; E. T. Bell, *Men of Mathematics* (New York: McGraw-Hill, 1937), pp. 424–25.

[28]Lermontova, p. 383.

Sofia did not assert herself, and only participated in class on rare occasions. Although she enjoyed her studies, she had little in common with her fellow students. According to the Russian biologist Kliment Timiriazev, who attended some courses with Kovalevskaia in Heidelberg, the demeanor of the German students toward her was "in general correct, but somewhat stupidly puzzled" by her desire to learn.[29] This did not bother Kovalevskaia, however. She was in Heidelberg to study.

The Kovalevskiis spent university vacations traveling together on Vladimir's geological forays, and on pleasure trips. They visited England, France, Italy, and different regions of Germany. They met with mathematicians, biologists, geologists, and other figures in the intellectual world. On a trip to England in early October 1869, Kovalevskii renewed his acquaintance with the naturalists Charles Darwin and Thomas Huxley, with whom he always had cordial relations.[30]

Huxley introduced Kovalevskaia to British mathematicians who in turn introduced Sofia to a wider circle of British intellectuals. It was through these contacts that Sofia was able to visit the novelist George Eliot at her salon.

For Eliot, the meeting was not an especially important event in her life, although she thought enough of it to note it in her journal for October 5, 1869:

> On Sunday, an interesting Russian pair came to see us,—M. and Mme. Kovilevsky [sic]: she, a pretty creature, with charm-

[29]Timiriazev, p. 24.

[30]Kovalevskii knew Darwin because of his publishing business. Darwin apparently liked Vladimir: he gave Kovalevskii the proofs of *Variation of Plants and Animals Under Domestication* before they were published, so that the Russian edition appeared in May 1867, while the original was delayed until January 1868. (Davitashvili, p. 70.) The novelist P. D. Boborykin also tells a story of the friendly relations of Darwin with Kovalevskii: "arriving in London, he immediately took himself off to Darwin, in whose family he was received always as a friend. And the first thing Darwin said to him after greetings was: 'I know now, who are you!' 'Who then?' asked Kovalevskii. 'A nihilist!' answered Darwin with laughter. In my article [*Fortnightly Review*, 1867] I mentioned the publishers of scientific and philosophic books of that tendency considered 'nihilistic'; I named Kovalevskii. And Darwin wanted to tease him about it." (Boborykin, vol. I, p. 506.) There is a packet of unpublished letters from Darwin to Vladimir in the Kovalevskaia papers of the IML. They indicate that Darwin had kind feelings for Kovalevskii, and that both Sofia and Vladimir visited Darwin in 1869.

ing modest voice and speech, who is studying mathematics (by allowance through the aid of Kirchhoff) at Heidelberg; he, amiable and intelligent, studying the concrete sciences apparently,—especially geology; and about to go to Vienna for six months for this purpose, leaving his wife at Heidelberg![31]

For Kovalevskaia, the Sundays at Eliot's salon were exciting, the more so since one day she had the opportunity to defend the rights and abilities of women against a formidable opponent: the philosopher and social-Darwinist theorist Herbert Spencer. Spencer believed that women's intellectual capabilities are vastly inferior to those of men. In fact, along with most social Darwinists, he was opposed to university education for women because he believed they would use up what little energy they possessed, and have none left for child-bearing.

Eliot instigated the argument without telling Sofia with whom she would be battling. She said only that here was a man who insisted that no woman was capable of scientific work. Kovalevskaia disputed with Spencer for three hours. She did not rout him, but she acquitted herself with distinction, and went away from the encounter with Eliot's praise to sweeten her failure: "You have defended our common cause well and bravely."[32]

While in Western Europe, Kovalevskaia worked for the cause of women's education and women's rights in general, just as she had in St. Petersburg. There she had been one of the four hundred women who signed the petition for women's higher education which Evgenia Konradi had circulated in late 1867. This act took a certain amount of courage, since the petition was presented to the government. Unfortunately, the petition had no immediate effect, so women wishing to study still had to go abroad.[33]

[31]J. W. Cross, *George Eliot's Life As Related in Her Letters and Journals* (Boston: Estes and Lauriat, 1895), vol. III, p. 78.

[32]Kovalevskaia, "Vospominaniia o Dzhorzhe Elliote [sic]," *VP 1974*, pp. 239–40.

[33]For the history of this petition, see Stites, *The Women's Liberation Movement in Russia*, pp. 75–6.

To accommodate such women, Sofia set up what her husband called a "women's commune" in Heidelberg. Membership in the commune was fluid. Its only constant participants were Iulia and Sofia. But at times their apartment housed Zhanna Evreinova, her sister Olga, and Aniuta. The revolutionary Natalia Armfeldt, a distant relative of the Korvin-Krukovskaia sisters, took refuge there as well.

Like Sofia, Aniuta, and Zhanna, Natalia Armfeldt was a general's daughter. She left her parents' home to study mathematics under Kovalevskaia's direction, but soon decided that active political agitation was more important than education. She returned to Russia to join the *Chaikovtsy*, a circle which brought political propaganda to the workers of Petersburg.[34]

Kovalevskaia tried to attract other women to her Heidelberg commune as well. She was concerned about Olga Lermontova, Iulia's cousin, and wrote to Vladimir: "From everything I hear about Olenka [Olga] it seems to me that she could become a very admirable woman; therefore we must use all our powers to draw her to us."[35]

Meanwhile, however, Kovalevskaia's relationship with Vladimir was beginning to deteriorate. In Lermontova's opinion, their marriage was adversely affected by two other members of the women's commune—Aniuta and Zhanna. Before their arrival, Iulia, Sofia and Vladimir had gotten along excellently together:

> The three of us would take walks in the vicinity of Heidelberg, go quite a distance and, finding ourselves on a flat stretch of road, the two of us, Sofa and I, would take off at a dead run, exactly like two little children. Dear heavens! How much gaiety and happiness there is in these memories of the first period of our university life!
>
> Sofa seemed to me then so happy—more than that, happy in a completely different way! . . . at that time [Vladimir] loved her with an entirely ideal love, without the slightest tinge of sensu-

[34]Lamanskii, TsGALI, f. 280, ed. khr. 340; E. K. Breshko-Breshkovskaia, "Iz vospominanii," *Golos minuvshego*, No. 10–12 (1918), pp. 192–207 and *passim*. Armfeldt eventually died in detention in Siberia.

[35]"Pis'ma S. V. Kovalevskoi," *Golos minuvshego*, No. 4 (1916), p. 85. Kovalevskii was in Vienna studying paleontological collections.

ality. She, apparently, related to him with the same type of tenderness. . . .

Only in that year [1869] do I remember Sofa as happy. A little later, even the very next year, it was already spoiled.[36]

When Zhanna arrived in Heidelberg at the end of 1869, Aniuta came from Paris to visit. Vladimir had to give up his room in the "women's commune" for the new arrivals, and took another apartment for himself nearby. Sofia visited him frequently, and the two took long walks alone or with Iulia. According to Lermontova, Kovalevskii rarely visited the women because Aniuta and Zhanna were often impolite to him. They believed that a fictitious marriage should be what its name implies: a formal legal device to facilitate travel abroad. They felt that Vladimir had fulfilled his role, and that now he should efface himself.

Aniuta's and Zhanna's attitude affected Sofia. She began to think that perhaps they were right, and that her relations with Vladimir were too close for a fictitious marriage. As Lermontova sadly remarks: "This interference of other people in the life of the young couple led to several quarrels and soon spoiled the good relations which had existed among the members of our little circle."[37]

Vladimir's letters to his brother during this period reflect his unhappiness with the new situation. There are ironic references to Zhanna, and he worries that he will lose such place as he had in his wife's thoughts and affections. But he does not see what the problem is, and blames Sofia's coldness on a personality quirk. In a letter of April 27, 1870 to his brother, Vladimir says: "In many respects she is a strange person, with whom it is necessary to be all the time, otherwise she becomes estranged."[38]

In the months following Aniuta's and Zhanna's appearance in Heidelberg, Sofia exhibited some of her least endearing character traits. She returned to the elder women's concept of fictitious marriage as far as the external courtesies went, and

[36]Lermontova, p. 382.

[37]*Ibid.*, p. 384.

[38]Letter in Shtraikh, *Sestry*, p. 134; see also *N N*, p. 224ff.

began to be rude and cold to Vladimir. But she had absorbed enough of his conception of the fictitious marriage to begin to take for granted his care of her, and she became capricious and demanding.[39] She saw a lessening of affection in his withdrawal after Aniuta's and Zhanna's arrival, and complained of his neglect.

Although Sofia herself was responsible for encouraging her husband to take up serious scientific work, she became jealous of the time he spent on field trips or immersed in books. Later, looking back on her relationship with Kovalevskii, Sofia said, unknowingly echoing Vladimir's letter to his brother: "He loved me only when I was near him. But he could always get along perfectly well without me."[40]

Neither Sofia nor Vladimir was capable of a reasoned evaluation of their relationship. Indeed, neither one of them was mature enough to be able to cope with the emotional ramifications of their marriage. Sofia was an inexperienced nineteen-year-old. Vladimir was eight years older, but his basic instability made him a poor mentor for the unfledged Sofia. However much the couple might have wanted to turn their marriage into a modern relationship *à la* the Bokovs or the characters in *What is to be Done?*, they had not the maturity and strength for the task.

Misunderstanding piled upon misunderstanding, as Sofia and Vladimir struggled with the complications caused by their unusual relationship. In her awkward attempts to get closer to Vladimir, Sofia began to make special bids for attention. She said she was afraid to stay in a hotel alone at night, so Vladimir had to give up the overnight field trips he had been intending to take during their vacation. She did not want to ride the train alone, so he had to accompany her or make sure Iulia or Zhanna was available to do so. She did not want to take time away from her studies to buy clothes (Kovalevskaia was notoriously unconcerned with clothes and personal appearance), so

---

[39]Interestingly enough, this capricious behavior was also characteristic of Olga Chernyshevskaia and Liudmila Shelgunova, two other wives whose husbands had planned modern, liberated relationships. See Bogdanovich, p. 40ff., for details.

[40]Lermontova, p. 384.

he had to select her outfits himself. Kovalevskii in his turn was willing to be imposed upon, even though he was now as involved in his geology and paleontology as Sofia was in her mathematics. But a store of resentment was building up. This tension augured ill for their future together.[41]

There were other difficulties as well. Vladimir's letters to his brother are filled with financial worries. The book business was a disaster, he was afraid his creditors would seize the brothers' small estate Shustianka, and he had to borrow three thousand rubles from General Korvin-Krukovskii to free Evdokimov after Zhanna's careless letter caused him to be arrested.[42]

Moreover, Vasilii Vasilevich soon found out that Aniuta was not in Heidelberg. In his anger at being deceived, he stopped sending her money. After this, Sofia had to divide the thousand rubles a year she received with her sister.

This sum would have been enough to support one person in modest comfort. When stretched to cover two, or even three or four, depending on how many stray women Sofia had taken under her wing, sometimes it simply was not enough. At one point Sofia had to turn down a request from Vladimir to help a young German woman he had met. She wrote to him that as much as she would have liked to help Vladimir's *protégé*, she simply could not manage it. Their apartment was overfull already, and they were so in need of funds that the commune members were thinking of hiring themselves out as cleaning women, she wrote to him in the winter of 1869–70.[43]

In the spring of 1870, Kovalevskaia journeyed to Paris to see Aniuta. After having studied diligently all winter, Sofia was looking forward to a week of intimate conversation with her sister. To her chagrin, as soon as Sofia arrived in Paris, Aniuta introduced her to Victor Jaclard, a handsome young Marxist

---

[41]Lermontova, pp. 384–86; letters in *N N*, p. 241; Shtraikh, *Sestry*, p. 134; Leffler, "Sonya Kovalevsky," p. 183. Elizaveta Litvinova had an interesting theory to account for Kovalevskii's docility. She felt that Vladimir encouraged his wife's foibles in order to bind her more closely to himself. (Litvinova, p. 46.)

[42]*N N*, pp. 230–32, 235, 241–42 and *passim*.

[43]Letter quoted in Shtraikh, *Sestry*, p. 126.

with whom she was living. According to Kovalevskaia's friend Maria Jankowska-Mendelson, when Sofia saw Jaclard:

> . . . it was exactly like a thunderbolt from the blue. What was this? During the time she had dreamed of how many secret societies, how many powerful organizations Aniuta had founded, how many great plans she had realized, how many Parisian workers had been subdued by her brave speeches, how many had been recruited for the family of revolutionary organizations—Aniuta had recruited only one! Some sort of medical student, a Frenchman no less![44]

Sofia felt betrayed. Aniuta and Zhanna had done their best to spoil her burgeoning relationship with Vladimir. Now, here was Aniuta, willingly submitting to what she had once called the tyranny of romantic love. This was not a development to which Sofia could easily reconcile herself.

Kovalevskaia's studies were going well. Her professors were pleased with her, and encouraged her to think that she was ready for mature graduate work in mathematics. But life outside of the classroom was complicated. Her marriage, which was neither fictitious enough to breed detachment, nor real enough to be emotionally satisfying, caused tensions. Increasing financial difficulties brought further anxieties, and Aniuta's love affair had caused a coolness between the sisters.

Moreover, the members of the women's commune were not all congenial to one another. Zhanna and Aniuta mocked Sofia's relationship with Vladimir, Iulia did not like Zhanna, Sofia and Iulia were impatient with those who would not apply themselves to study, and Aniuta's and Natalia Armfeldt's political views were more singlemindedly radical than those of the others.

It was therefore inevitable that the group would drift apart. It was as if they were a microcosm of the knowledge *vs.* revolution debate which shook the entire Russian radical movement

---

[44]Mendelson, p. 145. The affair was not legalized until more than a year later. But it was not uncommon for young nihilists not to formalize their marriages as a matter of principle. Kovalevskii's older brother, for example, did not marry his wife legally until the university at Kiev demanded it for his position. (Letter of 19 November 1869, *N N*, p. 222; Shtraikh, *Sem'ia*, p. 164.)

at this time.[45] Sofia, Iulia, and Zhanna decided to put education first. They told themselves that they would be able to help the common people later. Now, they were furthering the cause of women by obtaining their degrees. Aniuta and Natalia Armfeldt, on the other hand, opted for activism. Aniuta's writing and Natalia's mathematics could wait, they felt. The time to help the people was now.

The Heidelberg commune broke up during the spring and summer of 1870. Aniuta stayed with Victor Jaclard, pursuing political activity in France and fleeing with Victor to Switzerland when he was in danger of being arrested. Zhanna went to Leipzig to obtain her law degree. Natalia Armfeldt returned to Russia to conduct political agitation among the workers of St. Petersburg.

Sofia and Iulia went to Berlin, intending to concentrate on their studies. Both women came armed with excellent recommendations from their Heidelberg professors, and both were determined to complete their doctorates as soon as possible. And so they did, although the Paris Commune was to provide an exciting and dangerous interlude in the lives of Sofia and Vladimir Kovalevskii.

[45]See Meijer, p. 119ff. and *passim*, and Pomper, *Peter Lavrov, passim*, for details of this debate.

five

BERLIN AND THE PARIS COMMUNE

At the beginning of October 1870, Kovalevskaia arrived in Berlin. Her professors Leo Königsberger and Paul DuBois-Reymond had recommended to her the world-famous mathematician Karl Theodore Weierstrass (1815–1897), and she was eager to ask him to sponsor her at the university. She knew it would be difficult. Weierstrass had numerous students, was busy with administrative responsibilities, and was rumored to be opposed to the education of women.[1] But Sofia could be persistent, especially in connection with her mathematical career. She resolved to approach Weierstrass at his home.

The great mathematician later told Anna Carlotta Leffler that at their first meeting he had had no idea of Kovalevskaia's age or appearance, because she wore an unbecoming bonnet which hid her face and was suitable to a much older woman. When he heard her request to study with him, he was astounded. But Weierstrass was too understanding a man to turn anyone away without a hearing. He gave Sofia a series of problems and told her to come back when/if she could solve them.[2]

In a week Kovalevskaia returned with solutions in hand. Weierstrass sat her down and had her wait while he reviewed her answers. To his surprise, not only was everything correct, but the solutions were original and demonstrated Sofia's complete command of her field. Her paper showed him that she had "the gift of intuitive genius to a degree he had seldom

[1]Litvinova, pp. 34–5; Polubarinova-Kochina, *Zhizn'*, p. 14.

[2]Leffler, "Sonya Kovalevsky," p. 178.

found among even his older and more developed students," he later told Anna Carlotta Leffler.[3]

Leffler goes on to say that Weierstrass was Kovalevskaia's friend "from that hour," but in this, she was indulging in poetic exaggeration. In fact, the tsarist government had done a good job of slandering the women who went to Europe to obtain an education. The government spread rumors that all Russian women abroad were rabid revolutionaries, and immoral as well. It was even hinted that so many of them became medical students so that they could learn to perform abortions on their fellows.[4]

Given the reputation of Russian women abroad, it is not surprising that Weierstrass was wary because Sofia was Russian. To do him justice, though, he does not seem to have been worried solely on his own account. Rather, he was afraid that Kovalevskaia's nationality would prejudice the administration of Berlin University against her.

Soon after Sofia's second visit to him, Weierstrass wrote to her former adviser Professor Leo Königsberger in Heidelberg. He asked the latter to furnish him with detailed information on Sofia's mathematical level. Moreover, Weierstrass asked Königsberger whether "the lady's personality offers the necessary guarantees."[5] The Berlin mathematician needed this information for the faculty senate, which was to decide whether Sofia would be admitted to the university.

Königsberger's reply was satisfactory on all counts, and Weierstrass approached the faculty senate. He had the assistance of his colleague, the physiologist Emil DuBois-Reymond, whose brother, Sofia's Heidelberg mathematics professor, had written enthusiastically about his pupil's abilities. Weierstrass and Emil DuBois-Reymond were joined in their support of Kovalevskaia's application by the famous pathologist Rudolf Virchow and the renowned Hermann Helmholtz, another of Sofia's Heidelberg professors. But even the combined efforts of

[3]*Ibid.*, p. 179.

[4]See Meijer, p. 25ff. for discussion of the women's reputation.

[5]Karl Weierstrass, 25 October 1870 letter to Leo Königsberger, in "Briefe von K. Weierstrass an L. Königsberger," *Acta Mathematica*, 39 (1923), pp. 230–31.

four such eminent professors could not break down the age-old prejudice against women students in Berlin. The majority of professors, including the eminent mathematician E. E. Kummer, were opposed to Sofia's admission.[6]

Although the senate ruled against Kovalevskaia, Weierstrass could not conceive of abandoning someone of such outstanding mathematical gifts as Sofia. Despite his busy schedule, he proposed private lessons. He would recapitulate his lectures for her, and expound upon his own research. Kovalevskaia agreed eagerly. For the next four years she visited him every Sunday that she was in Berlin, and he in turn came to her one other day each week at the apartment she shared with Iulia Lermontova.

Kovalevskaia and Weierstrass had excellent rapport, both scientifically and personally. One of Weierstrass' biographers notes that "his role in both her scientific and personal affairs far transcended the usual teacher-student relationship. In her he found a 'refreshingly enthusiastic participant' in all his thoughts, and much that he had suspected or fumbled for became clear in his conversations with her."[7]

Weierstrass and Kovalevskaia started a correspondence which continued (with a three-year interruption on Kovalevskaia's side) until her death. Weierstrass' letters, which are the only ones to survive, are warm, loving, personal documents. Even when a letter is devoted to mathematics, it shows the strength of attachment between the older mathematician and the person he often called his favorite pupil.[8]

For Kovalevskaia and Lermontova, life was lonelier and even more devoted to study than it had been in Heidelberg. There,

[6]Lamanskii, TsGALI, f. 280, ed. khr. 340; Mittag-Leffler, "Weierstrass et Sonja Kowalewsky," p. 135; Litvinova, p. 34; E. El' [Elizaveta Litvinova], "Iz vremen moego studenchestva," *Zhenskoe delo*, April 1899, p. 52.

[7]Kurt R. Biermann, "Karl Theodore Wilhelm Weierstrass," *Dictionary of Scientific Biography* (New York: Scribner, 1973), vol. XIV, p. 223.

[8]Weierstrass' letters to Kovalevskaia are in the IML collection. They are published as *Briefe/Pis'ma von Karl Weierstrass an Sofia Kowalewskaja* (Moscow: Nauka, 1973); hereafter cited as *Briefe/Pis'ma*. Kovalevskaia's letters to Weierstrass, along with correspondence from other mathematicians and several mathematical manuscripts, were burned by him after Sofia's death.

they had had the other women of the commune, Vladimir Kovalevskii, and members of the Russian student colony with whom to while away the hours they could spare from their work. In Berlin, on the other hand, there was virtually no source of diversion. Kovalevskii was working on his dissertation at Jena University. Because of the combination of cool relations and lack of funds, he visited the two women rarely.

With no one to ensure that they rested from their studies, Sofia and Iulia would work sixteen or more hours a day, with only short breaks for poorly-cooked meals. Iulia at least had the daily exercise of a trip to her laboratory, but Sofia would refuse to leave her rooms for days on end. With little or no physical activity, and so much purely mental work, it is little wonder that Sofia slept badly, and grew thin and pale.[9]

One of the few social activities in which Sofia and Iulia engaged was visits to the Weierstrass family. The mathematician was a bachelor, and he lived with two unmarried sisters Clara (1823–1896) and Elise (1826–1898), who cared for his home.[10] The sisters loved Sofia, and were kind to Iulia as well. The three Weierstrasses, as Iulia recalled, "related to us warmly and tenderly, as if we were their children. They had Christmas trees for us, and invited us for dinner and for the evening. [But] they lived a solitary life, so we did not make any acquaintances at their house."[11]

According to Iulia, during the years in Berlin, "Sofa was always in the most unhappy frame of mind; nothing seemed to cheer her up, and she was always indifferent to everything but her studies. Her husband's visits always roused her a little, although the joy of meeting was frequently spoiled by mutual recriminations and misunderstandings."[12] Her friendship with Weierstrass and his sisters was pleasant, but Weierstrass was,

[9]Lermontova, pp. 378, 385–86.

[10]Bell, p. 409.

[11]Lermontova, p. 378. Affectionate letters from Clara and Elise Weierstrass to Kovalevskaia, and one letter from Sofia to Clara, are preserved in the Kovalevskaia papers of the IML.

[12]Lermontova, p. 385. In a note to this page, Shtraikh comments that there are several letters in the archives reflecting these disagreements between Kovalevskaia and Vladimir (p. 530, n. 385(1)).

after all, much older than she was. Moreover, he and his sisters could not understand why Sofia did not live with her husband.

At first Kovalevskaia did not attempt to explain the circumstances of her marriage to her adviser. As Litvinova pointed out, "one does not have to be a genius, but one certainly has to be a Russian, to understand that kind of thing."[13] Only in 1872, after almost two years of close mathematical contact with Weierstrass, did Sofia finally tell him that her marriage was not normal. It cannot be said that the German mathematician understood, and he certainly did not approve. But at least he realized that there were some extenuating circumstances for the couple's odd lifestyle.

Weierstrass' realization heralded a change in his attitude toward Sofia's mathematical studies. Previously, he had explained his research to her, and under his tutelage Sofia had mastered the most advanced techniques of analysis. But Weierstrass had seen little point in guiding her toward a dissertation. In his traditional view, married women did not need actual degrees because they did not need careers. Only in the fall of 1872, after Weierstrass had been informed of Kovalevskaia's true situation, did he consider that perhaps she would have some need of a career and therefore should work intensively toward her doctoral degree.[14] Meanwhile, however, events in France would necessitate a break in Sofia's mathematical studies.

As mentioned in the previous chapter, Aniuta accompanied Victor Jaclard to voluntary exile in Geneva. She joined the Russian section of the socialist First International Working Men's Association, mingled with other Russian and French political exiles, and earned money by writing children's stories for S. S. Kashperova's *Semeinye vechera* (Family Evenings). She also wrote to her parents, asking for her official documents so that she could legally marry Jaclard. Aniuta feared Victor's activities would lead to his arrest, and wanted to be able to help him. But

[13]Litvinova, p. 36.

[14]Weierstrass to Kovalevskaia, 26 October 1872, in *Briefe/Pis'ma*, p. 13.

since only legal wives had visiting privileges in French prisons, the pair needed to formalize their union.[15]

Victor and the other French political exiles kept a close watch on events in France. When it looked as if Napoleon III would be overthrown and a republican government had a chance of success, they secretly returned to their homeland. In September 1870, Jaclard and Aniuta were in Lyons just as the republic was declared. He was elected as one of three Lyons representatives to the new assembly in Paris. In October he was arrested for taking part in Louis Blanqui's attempt to overthrow the conservative leadership of the republic. Jaclard was detained, but acquitted of all charges in March 1871, possibly because he had shown himself to be popular among the common people of Paris.[16]

Meanwhile, Bismarck's strategy in the Franco-Prussian War had been successful, and the well-equipped Prussian army was advancing rapidly on Paris. The new premier Adolphe Thiers fled to Versailles with members of his government and a large body of troops, leaving only the National Guard to defend the city. When Thiers made a humiliating peace with the Germans and tried to regain control of Paris, the National Guard refused to surrender their weapons and on March 18, 1871 declared the Commune.

Much has been written about the Paris Commune of 1871. It has been attacked or praised from every conceivable point of view. Suffice it to say that political radicals and socialists saw the Commune as a significant harbinger of the world-wide revolution to come. Marx and Engels analyzed it extensively, and Lenin said: "The Commune is the first attempt by the proletarian revolution to *smash* the bourgeois state machine . . ."[17]

[15]Letters to Sofia from Aniuta, *VIP 1951*, pp. 333–36.

[16]Knizhnik-Vetrov, *Russkie deiatel'nitsy*, pp. 180–85. In February elections for the new assembly in Paris, Jaclard received 60,000 votes (as opposed to 52,000 for Blanqui). Jaclard was one of the few leaders of the French socialist movement at this time to be an avowed follower of Marx.

[17]V. I. Lenin, *Gosudarstvo i revoliutsiia* (State and Revolution, 1917) (Moscow: Politizdat, 1973), p. 57. For other views of the Paris Commune, see Karl Marx, "The Civil War in France," and his and F. Engels' additions to this work; and *The Paris Commune 1871*, Roger Williams, ed. (New York: Wiley, 1969).

Both Aniuta and Victor took part in the events and decisions of the Commune. Aniuta was one of five people working on education, and had special responsibility for women's education. She and the French feminist-socialist André Leo founded a newspaper *La Sociale*, which was published daily during most of the few weeks of the Commune's existence. Jaclard was a member of the Paris Committee of Public Safety, and commander of the Montmartre National Guard unit as well.[18]

The Kovalevskiis had been intending to visit Paris for scientific reasons, but Aniuta discouraged them because of the Franco-Prussian War and impending internal upheaval.[19] The news from Paris was so ominous, however, and Kovalevskaia was so worried about the lack of letters from her sister, that the couple decided to risk the trip.

They could not get a pass through the German army of occupation which surrounded the city, so they walked the shores of the Seine until they found a small abandoned boat. They got in and started rowing. Challenged by a sentry, they ignored the threat and rowed more quickly. There was some gunfire, but it was dark, and the sentries' aim was bad. Sofia and Vladimir managed to make their way into Paris unhurt.[20]

Kovalevskii later wrote to his brother that "from April 5 to May 12 we lived happily in the Commune in Paris."[21] Sofia became involved in hospital work; even the small amount of medical training she had received in St. Petersburg was appreciated by the desperate Parisians. She and Aniuta nursed those wounded in the Versailles government's bombardment, which had started April 2. Other Russian radical women were acting as nurses as well. Sofia and Aniuta encountered several women who had been in their circle in Russia.[22]

Meanwhile, Vladimir prowled the Paris paleontological collections, and talked to such specialists as had not fled the city.

[18]Knizhnik-Vetrov, *Russkie deiatel'nitsy*, pp. 186–90; Shtraikh, *Sem'ia*, p. 188.

[19]See Aniuta's letters in Shtraikh, *Sestry*, p. 143ff. and *VIP 1951*, p. 333ff. As a rule, Shtraikh's quotations are less abridged than those in *VIP 1951*.

[20]Leffler, "Sonya Kovalevsky," pp. 184–85.

[21]9 August [1871] letter in *N N*, p. 255.

[22]Leffler, "Sonya Kovalevsky," p. 185.

The cannonfire does not seem to have bothered him. While the bombs were literally bursting over his head, he came to the conclusion that his main area of study would be fossil mammals.[23]

After six weeks in Paris, Kovalevskaia longed to return to her mathematics in Berlin. She left with a clear conscience, since Aniuta had urged her to go, and everyone felt that the city could withstand the siege for several more months. On May 22, however, Paris fell to the Thiers forces. A reaction began, during which over 10,000 men, women, and children were shot down by the armies of MacMahon and Galifet.

As soon as she heard of the fall of Paris, Sofia began to worry about her sister. When news came that Jaclard had been arrested, Sofia and Vladimir prepared to leave Berlin.[24] A rumor that Aniuta also had been taken hastened their departure. By the tenth of June, the Kovalevskiis were back in Paris, which Vladimir was of the opinion they should never have left in the first place.

To their relief, they found Aniuta safe, although her comrade André Leo had been arrested, and Aniuta was being sought by the government. The Kovalevskiis spirited Aniuta away to London, where Marx helped her to establish herself temporarily.[25] Sofia and Vladimir sent an anxious message to General Korvin-Krukovskii at Palibino, and remained in Paris to see if they could do anything for Jaclard.[26]

While they waited, Vladimir evolved fantastic plans for the common future of himself, Jaclard, Aniuta, Sofia, and even his unsuspecting brother Aleksander. Jaclard would probably be sent to the New Caledonian penal colony, Vladimir surmised. He decided that he would accompany Aniuta to New Caledonia, while Sofia would stay behind to finish her studies. He

[23]Borisiak, p. 24.

[24]During the first few days of reaction, several people were mistaken for Jaclard and shot out of hand, but by the time he was taken, the Thiers government had decided to have show trials, so he was imprisoned awaiting judgment. (Knizhnik-Vetrov, *Russkie deiatel'nitsy*, p. 191.)

[25]*Ibid.*, p. 195.

[26]V. O. to A. O. Kovalevskii, letters of 11 July and 9 August 1871, *N N*, pp. 399 and 255, respectively.

said nonchalantly that he would get all the money he needed from General Korvin-Krukovskii, and everyone would settle happily in New Caledonia. In addition, he blithely informed his brother that when Sofia had finished her work, Aleksander could escort her to her new home in the South Pacific.

Vladimir's justification for the upheaval he was suggesting for himself and his brother was simple. "Sofa and Aniuta have become so dear to me that I could not part with them forever," he wrote to Aleksander.[27]

This elaborate plan was typical of Vladimir, and illustrates his failings as well as his virtues. He was generous with his own time and resources, and regarded no inconvenience to himself as too great in the service of those he loved. But Vladimir tended to be generous and not overscrupulous about using the time and resources of others as well.[28] He was impetuous, rarely thinking out anything to its logical conclusion. Moreover, he was given to statements of affection or loathing which he would promptly contradict in just as extreme a manner; hence his expressed devotion to Aniuta, after months of antagonism toward her.

As it happened, the sacrifices that Vladimir anticipated for himself and his brother were not needed. Vasilii Vasilevich and Elizaveta Korvin-Krukovskii arrived in Paris at the beginning of July, and set about working on Jaclard's release. The general was armed with a letter from one of his diplomatic friends to the Russian ambassador, which said: "Help, I ask you, the unhappy old man . . . do not refuse your support."[29] Moreover, Vasilii Vasilevich knew Thiers slightly, and was prepared to trade on the acquaintance to help Aniuta's Communard lover.

There are almost as many theories about Jaclard's escape as there are historians and memoirists who have written about the

[27]11 June 1871 letter to Aleksander, in Shtraikh, *Sestry*, p. 169. Parts of this letter appear in *N N* and *VIP 1951*, but Shtraikh is most complete.

[28]Several times during his career as publisher and later financial speculator, Vladimir appropriated his wife's, brother's, or Iulia Lermontova's funds. In all cases, he promptly lost the funds he had acquired by staking them on some get-rich-quick scheme.

[29]Quoted in Knizhnik-Vetrov, *Russkie deiatel'nitsy*, p. 195.

event. Anna Carlotta Leffler and the Polish revolutionary Maria Jankowska-Mendelson, both of whom were friends of Kovalevskaia and talked with her about the incident, feel that General Korvin-Krukovskii's talks with Thiers were of prime importance in engineering the escape. According to them, Thiers received Vasilii Vasilevich cordially, said he could not procure a pardon for Jaclard, but casually mentioned that Jaclard was to be transferred to another prison the next day. He also commented that the prison procession would pass by a building where a large exhibition was being held.[30]

The next day, as the prisoners moved through the crowded streets, a woman grabbed Jaclard by the arm and vanished with him into the throng in front of the exhibition hall. Leffler says the woman was Aniuta, although it must have been Sofia if this version of the escape is correct, since Aniuta was already in London. Knizhnik-Vetrov offers two alternatives: Jaclard's sister or perhaps Sofia. Shtraikh feels that Vladimir was the one to seize Jaclard. Nechkina maintains that Kovalevskaia herself obscured her own role for fear that rumors would get back to the wrong people.[31] Leffler's story, however, with Nechkina's substitution of Sofia for Aniuta, is probably correct in all essential details.

Whether or not the Kovalevskiis were involved in Jaclard's escape from the prison itself, it is clear that they helped him flee France. Vladimir gave Victor his passport, thereby enabling the Communard to travel to Switzerland.[32] Aniuta and the elder Korvin-Krukovskiis joined him there.

Maria Jankowska-Mendelson explained the general's attitude toward Jaclard in a curious manner: "He was not bothered by the Communard. As a military man, he obviously sympathized with the wrath of Paris against those who had signed the Versailles peace."[33] In any case, Vasilii Vasilevich finally made his peace with Aniuta and her beloved. Victor's

[30]Leffler, "Sonya Kovalevsky," p. 187; [Jankowska] Mendelson, p. 146.

[31]Knizhnik-Vetrov, *Russkie deiatel'nitsy*, p. 195; Shtraikh, *Sem'ia*, p. 194; Nechkina, p. 87. Nechkina downplays Thiers' role, saying that historians know him too well to credit him with such kindness.

[32]Winter 1871 letter from Aniuta to Vladimir, *VIP 1951*, p. 338.

[33]Mendelson, p. 146.

and Aniuta's marriage was legalized in the presence of the general and his wife.

Soviet historians frequently speculate on whether the visits of Sofia and Vladimir to Paris both during and after the Commune were scientifically or politically motivated. Shtraikh and Davitashvili seem to think that the pair absentmindedly wandered into Paris in search of mathematicians and rocks, and either forgot about or paid no heed to the German blockade and the Versailles bombardment. Nechkina, on the other hand, is impatient of purely scientific motivations for the couple's six-week stay in the Commune:

> These proposals seem naive. The Kovalevskiis could undoubtedly have found some other time for scientific meetings and work in Paris, so one cannot doubt that they were drawn to the revolutionary city not only by the interests of 'abstract science'. Of course, anxiety for Sofia Kovalevskaia's sister . . . drew them to Paris; but in my opinion even this circumstance cannot settle the question of the reasons for the Kovalevskiis' trip. The reasons for their appearance in the revolutionary city were more complicated: a deep sympathy for the revolutionary struggle was among the forces which directed their actions . . . [34]

If either of the two could be thought to have gone to Paris purely for the sake of science, it would have to be Vladimir. Yet his letters to his brother both before and after the Commune do reflect, as Nechkina claims, a "deep sympathy" for the struggles of the red republicans and later the Communards. In an August 18, 1870 letter he cheerfully mused on the possibilities of a French revolution if the Germans took Paris, and said, "in general the times are as interesting as they can possibly be." Two days later, he wrote: "I pray for one thing: that the Prussians will beat the French so badly that the latter will overthrow Napoleon, and then the republican troops can destroy the Prussians."[35]

Throughout the events of the Commune, Kovalevskii remained on the side of the rebels, and he mourned them with

[34]See Shtraikh, *Sestry*, p. 145; Shtraikh, *Sem'ia*, pp. 185, 189; Davitashvili, p. 53; Nechkina, pp. 82–4, her quotation is on page 84.

[35]Both letters in Shtraikh, *Sestry*, pp. 145, 148.

sincerity. In a May 28, 1871 letter to his brother, he called the reprisals far worse than the notoriously bloody June Days in the revolution of 1848, and lamented that "very many of our good friends have been wounded or killed . . . The best and [most] energetic people were shot down everywhere . . ." Further, he told Aleksander that he could not blame the Communards for destroying public buildings and executing hostages. He would have done the same in their place:

> They had two hundred hostages, and because the Versailles troops were shooting everyone, the insurgents, not having achieved either an exchange [of prisoners for the hostages] or an amnesty, shot sixty-three people, including the archbishop . . . . Of course, the Versailles forces could have rescued them easily, but they deliberately did not, so that all the *odium* of the killing of hostages would fall on the insurgents.[36]

Unfortunately, there is little record of Kovalevskaia's thoughts on the Paris Commune. Unlike Vladimir, she did not write long letters to anyone during this period of her life. Her closest confidante was probably Lermontova, since Aniuta's relationship with Jaclard seems to have ended the intimacy of the sisters. But Sofia lived with Iulia in Berlin, so there was no need to write in detail to her friend.

It is likely, however, that Kovalevskaia's opinions of the Commune were the same as those of her husband. She was not as involved in political agitation as her sister was, but she admired those who were, and was herself willing to take risks on occasion. Sofia planned to write the story of her weeks in the Paris Commune, and meant to portray the experience positively, but like so many of her literary projects, this one seems to have been unrealized at the time of her early death.[37]

It would be a mistake to characterize Sofia as a revolutionary activist. But it would also be a mistake to deny that she was motivated by political considerations. She was committed to

[36]Letter in *N N*, pp. 251–52.

[37]Leffler, "Sonya Kovalevsky," p. 185. There is a slight possibility that a manuscript may yet turn up in the Soviet or Swedish archives. L. A. Vorontsova found Kovalevskaia's unfinished novel about the Chernyshevskiis only in 1953!

the furthering of women's rights, had a reputation as a leader in the women's movement, and was conscious of her importance as "a shining light toward which the eyes of all young girls who wanted an education turned," as Elizaveta Litvinova put it.[38]

Although she was not a full-time revolutionary herself, Kovalevskaia had ties with populist and socialist circles both in Russia and abroad. Her friends included the Chernyshevskii family, Kropotkin, Lavrov, Jankowska-Mendelson, several former Communards, and many others. As will be seen, she often performed services for these people, especially the women, sometimes putting herself in danger as a result.

Of course, for most of her adult life, Kovalevskaia's main intellectual activity was mathematics. Yet with the exception of the years in Berlin, Sofia seldom devoted herself entirely to that subject. She was too involved in the political and social movements of her day, and had too complicated a personal life, to be able to devote as much time to her beloved mathematics as she would have liked.

Even in Berlin, her concentration was broken by friction with Vladimir. In fact, this bickering was to reach such proportions that it interfered with her prospects of finishing her degree. Only the temporary settlement of their conflicts would allow them to pursue their studies with relative peace of mind, and enable Kovalevskaia to obtain her doctorate.

---

[38]E. El' [Elizaveta Litvinova], "Iz vremen moego studenchestva," p. 34.

six

A DOCTORATE IN MATHEMATICS

As autumn of 1871 approached, Kovalevskaia was eager to return to Berlin to continue her mathematical studies. She had spent the spring tending the wounded in a hospital in the Paris Commune, and the summer worrying about Jaclard's escape and encouraging Aniuta's resumption of good relations with her parents.

Back in Berlin, Sofia and Iulia Lermontova, who had gone to Russia for the summer vacation, settled into their old routine of work, work, and more work. Soon, it must have seemed to Sofia as if the events of the past five months had never happened.

Kovalevskaia resumed the exchange of weekly visits with Weierstrass. He would go over his university lectures with her, and as time went on and Kovalevskaia became more mathematically sophisticated, Weierstrass would discuss his current research as well. Sofia valued these meetings highly: "They had an extremely important influence on my whole mathematical career. They finally and irrevocably defined the direction which I followed in my future scientific work. All my works are done precisely in the spirit of Weierstrass' ideas."[1]

Weierstrass seems to have valued these meetings as highly as did Sofia. Weierstrass was the greatest mathematical analyst in the world; "the master of us all," as the famous French mathematician Charles Hermite called him. He numbered among his students and disciples some of the best mathemati-

---

[1]"Avtobiograficheskii rasskaz," *VP 1974*, p. 371.

cians of the last century—Leo Königsberger, Kurt Hensel, Hermann Schwarz and many others. But he often said that Sofia was his most talented and favorite pupil. She understood his system of analysis as well as anyone except Weierstrass himself, and was often called upon to explain his work to other mathematicians.[2]

Kovalevskaia's personal life was not going as smoothly as her mathematical studies. During the winter of 1871–72, Vladimir Kovalevskii began to think in terms of divorce. A few months before, he had written to his brother that he could not bear to be parted from Sofia and Aniuta, and would follow them to New Caledonia. Now, in what was for him a typical reversal of feeling, he decided that he did not want or need a wife who was intellectual, independent, and admittedly difficult to live with. Above all, he no longer wanted a fictitious wife.

Vladimir deliberately began to instigate quarrels. He told himself and his brother that Sofia would make a horrible wife and mother, and made allusions to obtaining his freedom. Kovalevskii poured out his feelings in a November 1871 letter to his brother. He said he had affection for Sofia, although he was not now and never had been in love with her. He felt that Sofia would be a bad mother, and he a bad father. He agreed with people who said that he and Sofia were unsuited to one another, and said "we both now repent of our marriage." He was even willing to divorce Sofia, he told his brother.[3]

Kovalevskii enlisted Aniuta's aid in asking Sofia about ending their marriage. Aniuta attempted to talk to her sister, but had no success. She wrote to Vladimir that Sofia was in "some kind of abnormally passive state of mind" and would make no decisions.[4] Kovalevskii wrote to his wife directly, but his hints

[2]Mittag-Leffler, "Weierstrass et Sonja Kowalewsky," pp. 133, 172; Heinrich Behnke, "Karl Weierstrass und seine Schule," in *Festschrift zur Gedächtnisfeier für Karl Weierstrass 1815–1965* (Köln: Westdeutscher, 1966), pp. 13–40; Hermite, 27 February 1882 letter to Kovalevskaia, IML.

[3]Quoted in Shtraikh, *Sestry*, pp. 175–76.

[4]Quoted in Davitashvili, p. 59.

were so ill-phrased and disingenuous that Kovalevskaia responded with an irritated, formal reply:

> I have just received your letter, and I won't stop to say how much it pained me, because it was obviously written with just that intention. But you are entirely mistaken if you think that I have any 'orders' or 'instructions' to give you. I think that it is completely superfluous to tell you that it could never enter my head to make use of the noble offers which you tacitly hint at in your last letters . . .[5]

In the next two years the relationship of Vladimir and Sofia alternated between periods of hostility and affection. Kovalevskii received his doctorate from Jena University in Germany in March 1872, with a dissertation in paleontology that was widely acclaimed in scientific circles. He then traveled to London to see Darwin, Huxley, and other naturalists. While there, he became fascinated with the idea of a "happy home" on what he thought was the English model: "One only has to live a while in England to understand all the joy of family life, especially with such delightful people as Englishwomen. But it's better for me not to think about it, at least until we can free ourselves legally—although I still don't know how to do it."[6]

In spite of his frequent mentions of divorce in his letters to his brother, Vladimir does not seem to have broached the subject to his wife again.[7] Instead, he inexplicably decided to fall in with a plan of Aniuta to unite him with her sister.

Aniuta and Victor Jaclard had moved to Zurich, and Aniuta was expecting a baby. Since women could enroll at the university and Polytechnicum in Zurich, Aniuta decided to invite Sofia to live with her. Sofia could attend the Polytechnicum for a semester or two and thus widen her knowledge of physics and mechanics. Aniuta hoped that Vladimir would visit fre-

---

[5]Quoted in Vorontsova, p. 127. Kovalevskaia goes on to say that she curtails her freedom less than Vladimir thinks, but this seems to have been nothing more than a flip remark designed to irritate Vladimir.

[6]Letter to Aleksander, quoted in Shtraikh, *Sestry*, p. 193.

[7]See letters in *ibid.*, pp. 174–93 and *passim*; *N N*, p. 298; Davitashvili, p. 56ff.

quently and, in the congenial family atmosphere of the Jaclard apartment, Sofia and Vladimir would come together.

Aniuta also had another, rather surprising reason for wanting to draw her sister away from Berlin. She felt that Sofia was becoming too absorbed with Weierstrass. This "exclusive absorption" was unhealthy, Aniuta felt. Sofia needed to broaden her scientific background to include other schools and methods, Aniuta wrote to Vladimir.[8]

The sentiments expressed in Aniuta's letter could be interpreted to mean that Aniuta had heard and possibly even believed the gossip circulating in Russian colonies abroad about Kovalevskaia and Weierstrass. Vladimir had heard the stories and amusedly discounted them. He wrote to his brother: "I greatly enjoyed the story about Sofa. I can only add that it certainly would have happened in fact, if the professor 'with whom she lives' (she sees him twice a week) were not seventy-two or thereabouts."[9] Aleksander Kovalevskii also disbelieved the rumors, although he did advise his brother to ask Sofia just how long one needed to study a single subject in order to master it.[10]

All observers close to the situation agreed that Weierstrass' emotion toward Sofia was of the fatherly rather than the connubial sort. Even Anna Carlotta Leffler, who was always eager to draw portraits of her friend's romantic attachments, felt that Weierstrass loved Sofia only as a daughter.[11]

In some sense, it is a pity that there was no sexual relationship between Kovalevskaia and Weierstrass. Had there been less of a difference in age (Sofia was fully thirty-five years younger than her professor) and had Sofia not been married (Weierstrass respected conventional ties), the case might have been otherwise. The two had much in common and in many ways would have suited each other, in spite of their different nationalities. Certainly Weierstrass was a much more stable personality than Vladimir, and his staidness could have benefited from Kovalevskaia's high spirits.

[8]Letter to Vladimir, quoted in Shtraikh, *Sestry*, p. 190.

[9]*N N*, p. 305. Weierstrass was really fifty-seven.

[10]Shtraikh, *Sem'ia*, p. 244.

[11]Leffler, "Sonya Kovalevsky," p. 178ff.

The rumors of an affair between Weierstrass and Kovalevskaia have been discussed in some detail, not because there is any doubt that the gossip was false, but because shadows of these and other rumors crop up repeatedly.[12] Kovalevskaia was an attractive, articulate, political woman. She succeeded as a professional in a field which, until her advent, had been almost exclusively male. It is not surprising that she excited rancor among men jealous of their prerogatives as the intellectually superior sex. As Weierstrass' biographer Kurt Biermann put it: "Her links with socialist circles, her literary career as author of novels, and her advocacy of the emancipation of women strongly biased the judgment of her contemporaries, and resulted in the defamation of the friendship."[13]

One should add to Biermann's list Kovalevskaia's position as the first woman to obtain her doctorate in mathematics, her professorship, and her other academic honors. It is easy to see why Sofia excited envy among her contemporaries. Rumors would pursue her throughout her life.

Aniuta's plans to unite Vladimir and Sofia were not put into action until the spring of 1873, when Aniuta was about to give birth to her baby. There was ample excuse to ask Sofia to visit, since the Korvin-Krukovskii parents had come from Palibino to await the birth of their first grandchild. However, to ensure Sofia's favorable response to an invitation, Aniuta made efforts to become acquainted with Elizaveta Litvinova. Litvinova was studying mathematics in Zurich under a former student of Weierstrass, Professor Hermann Schwarz. Aniuta clearly hoped that Elizaveta could entice Kovalevskaia to stay in Zurich with promises of comradeship in their mathematical studies.[14]

Litvinova had long been aware of Sofia's existence. She had first heard of her through Evgenia Konradi, who had collected

[12]Even now, the gossip has not ceased. The Venezuelan mathematician and historian of science Francisco Jose Duarte, for example, goes so far as to castigate other biographers of Kovalevskaia and Weierstrass for not giving this supposed romance enough play in their accounts of the pair. (*Biografias* [Caracas: Academia de Ciencias, 1969], p. 140.)

[13]Biermann, p. 223.

[14]E. El', "Iz vremen," p. 37.

signatures on a petition for women's higher education in St. Petersburg in the winter of 1867–68. Konradi and another staunch worker for women's rights, Nadezhda Stasova, had helped Litvinova to arrange calculus and trigonometry lessons for herself. Stasova allowed Litvinova to use her apartment for sessions with A. N. Strannoliubskii, who had been Kovalevskaia's first teacher of higher mathematics. Thus, Elizaveta had heard much about Sofia. She considered Kovalevskaia a model worthy of emulation, and was eager to meet her and persuade her to stay in Zurich.[15]

Unfortunately, Aniuta's plans with regard to Vladimir did not go so smoothly. Earlier that spring, he had journeyed to Odessa for what he had considered would be the mere formality of his magistral examination.[16] In his usual fashion, he had not listened to the warnings of his brother or his friend Ivan Sechenov, and had gone into the exam unprepared. Moreover, he had publicly insulted the work of one of the chief examiners not long before his examination. After two question sessions and acrimonious debate with the offended examiner, Kovalevskii failed to obtain the necessary vote for his degree.

Kovalevskii's biographers all agree that to a large extent he brought the disaster upon himself with his overconfidence and rash criticism of the scientific abilities of his examiner.[17] But Vladimir's examination was also affected by circumstances beyond his control. At that time, the Russian scientific community was just beginning to shake off the domination of mostly German foreign specialists in the Academy of Sciences and university system. Consequently, Russian scientists tended to mistrust Russians who had received all their training in Ger-

[15]*Ibid.*, pp. 34–6; Stasov, p. 240.

[16]The Russian system of degrees was and still is different from the American system. A magistral degree (Russian master's degree) was needed for university positions, and a Russian doctorate was required for high university and Academy of Sciences posts. The magistral degree was approximately equal to an American Ph.D. today. The closest equivalent to a Russian doctorate in an American university is probably a full professorship or a prestigious chair.

[17]Shtraikh, *Sem'ia*, p. 232ff.; Shtraikh, *Sestry*, p. 180ff. (with extensive letter quotations about Vladimir's confidence beforehand, and his despair afterward); Borisiak, pp. 34–5; Davitashvili, pp. 165–77.

man institutions, because they felt that the foreign-trained Russians would side with the remaining Germans on questions of academic politics and promotions. This mistrust hurt Vladimir. Later, it seems to have played a role in the reception of Kovalevskaia in St. Petersburg as well.[18]

Vladimir was aware of these undercurrents in the Russian scientific world. But he sided fully with the Germans in the Russian Academy, and was arrogant and overbearing in his behavior toward most of his Russian colleagues. When he failed his exam, Vladimir refused to recognize that he was in any way at fault. His reputation in the Russian scientific community was hurt by the Odessa fiasco, and he damaged it further by ranting against anyone connected with the examination.[19]

Kovalevskaia at first knew nothing of her husband's disgrace in Odessa, and was irritated that she did not find him awaiting her in Zurich. She wrote him from there in April, in a cold, lofty tone, pointing out that his absence put her in an uncomfortable position. Her father was beginning to comment on how little time Sofia and Vladimir spent together, she wrote. Her parents suspected that there was something odd about the marriage.

Kovalevskaia went on to say that she had not wanted to see Vladimir in Zurich at first, because they would be under the sharp eyes of her parents. But after talking with Aniuta and reading his letters to her, Sofia had come to the conclusion that he was as unhappy as she was. Sofia hinted that she might be willing to consider ways "to make family life possible" for Vladimir by agreeing to a divorce. She invited him to visit her in Berlin in May:

> After I have talked this over with you I will be able to devote myself to my studies with a quieter mind; after I have made some kind of decision, I will be much calmer. Now, all these interrogations, hints, pestering, this unclarity about my future

[18]For a description of these hostilities in the Russian scientific community, see Alexander Vucinich, *Science in Russian Culture 1861–1917* (Stanford: Stanford University Press, 1970), chapters 1–3 and *passim*.

[19]Vladimir Kovalevskii, *Zametka o moem maqisterskom ekzamene* (Kiev: Universitetskaia Tipografiia, 1874).

frequently upset me greatly. But of course, in that respect, do whatever is more convenient for you.[20]

For a short while that spring, then, it looked as if a divorce had been agreed upon, at least in principle, by both parties. Almost immediately, however, both Sofia and Vladimir had second thoughts. After Sofia heard about Vladimir's humiliation in Odessa, she forgot about his recent slights in her desire to comfort him. Vladimir, in turn, when faced with the real possibility of obtaining the divorce he had thought he wanted, backed off, and apologized to Kovalevskaia for the "malicious letter" he had sent her.[21]

There followed a month's exchange of increasingly affectionate letters, including one in which Sofia half seriously, half gaily refers to the gossip about her and Weierstrass. "In my new friendship there is much that is poetic, idealistic, and sincere," she wrote to Vladimir. "It gives me incredibly much happiness and enjoyment, but alas, there is nothing romantic in it at all!"[22]

Vladimir and Sofia met in Berlin and Zurich, and resumed their previous friendship. Elizaveta Litvinova, who saw the couple frequently in late spring and early summer, found them cordial with one another, although Sofia seemed to be passive. Litvinova said that her Russian student friends at the dormitory could not understand the Kovalevskii marriage. All knew it to be fictitious, since Kovalevskaia was well-known among the young Russian women of Zurich. Yet Sofia and Vladimir seemed "such a tender couple."[23]

Vladimir's letters to his brother began to mention "Sofa" frequently and with tenderness. For the past year, she had been almost absent from his letters to Aleksander. Now, he reverted from his formal references to "Sofia Vasilevna," and

[20]Spring 1873 letter in Shtraikh, *Sestry*, pp. 197–98. It is interesting that both Kovalevskaia and her husband have a Dostoevskian way of being polite and obliging when they are most angry.

[21]2 May 1873 letter in *ibid.*, p. 199.

[22]Spring 1873 letter quoted in *idem*.

[23]E. El', "Iz vremen," pp. 47–8.

began calling her "Sofa" again. Soon, Vladimir notified his brother that he was moving to Berlin to be with her.[24]

The summer of 1873 saw the renewal of affectionate relations between Sofia and her husband. From then on they traveled together, took rooms near one another in Berlin, and wrote joint letters to Aleksander Kovalevskii. More significantly, they began to make common plans for their future in St. Petersburg.[25]

Renewed cordiality made scientific work much easier for both Sofia and Vladimir. Vladimir was able to devote all his attention to the fossil collections he was examining that summer. His work during this period was praised and cited by paleontologists throughout Europe and America.

Sofia concentrated on making a second attempt to produce a doctoral dissertation. She had already prepared one paper, which Weierstrass had approved and recommended for publication. Before she could submit it to a journal, however, she was forestalled. The young Zurich professor Hermann Schwarz sent a preprint to Weierstrass which contained essentially the same results as she had obtained, so it was necessary to begin again.[26]

Sofia was disappointed, but such coincidences of discovery are not uncommon in mathematics. While she was in Zurich that spring and summer, she had several long discussions with Schwarz. Kovalevskaia was so interested in Schwarz's work, and so charmed with him when they met in Zurich, that she seriously considered leaving Berlin to study with him. Surprisingly enough, it was Aniuta who opposed this plan, saying it would hurt Weierstrass too much.[27]

Kovalevskaia returned to Berlin, and worked diligently through the fall and winter of 1873. By the spring of 1874, she had written three papers, "each one of which deserved a doc-

[24]See 1873 letters in *N N*, pp. 319–29, *passim*.

[25]Two of Kovalevskii's biographers feel that the cordiality of Sofia's and Vladimir's feelings extended to sexual relations at this time. (See Borisiak, p. 38; Davitashvili, p. 60.) The couple's letters, however, do not seem to support this view.

[26]E. El', "Iz vremen," p. 39.

[27]*Ibid.*, p. 51.

toral degree."[28] The first was without question the most important. It has been called "the point of departure" for future
research on the subject and is "now regarded as the first
significant result in the general theory of partial differential
equations."[29] The second paper dealt with the form of Saturn's
rings, which was a classical problem in theoretical astronomy.
The third was concerned with reducing certain classes of integrals to less complicated forms. All three works will be discussed in more detail in a later chapter.

Weierstrass and Kovalevskaia had become so absorbed in the
production of these papers that neither one of them considered
exactly where they would present them when they were done.
They decided to try Göttingen, because Weierstrass knew
mathematicians there, and because the university had a reputation for granting degrees to foreigners *in absentia*.[30] There was
an exchange of letters on bureaucratic matters during June and
July.[31]

At times, it looked as if the cause was hopeless. One problem
was that Weierstrass wanted to spare Kovalevskaia the ordeal
of an oral examination if at all possible. The reason for this is
not entirely clear. Perhaps he thought that Sofia's German was
not good enough. Possibly, he thought that she would be subjected to unpleasantness by her male examiners.

In any case, Weierstrass had Kovalevskaia write an obsequi-
ous letter to the Dean of the Göttingen Philosophical Faculty,
which said in part:

> . . . I hope the very reverend Dean will not misconstrue me if
> I acknowledge openly that I do not know whether I have

[28]Litvinova, p. 38.

[29]Charles Hermite, 21 May 1889 letter of support for Kovalevskaia's profes-
sorship at Stockholm, Kovalevskaia papers, IML; P. Ia. Polubarinova-
Kochina, "On the Scientific Work of Sofya Kovalevskaya," translated by Neal
Koblitz, in Kovalevskaya, *Russian Childhood*, p. 231.

[30]There was precedent for awarding the degree to a woman. Dorothea
Schlözer Rodde, daughter of a German historian in the Russian Academy of
Sciences, received such a degree in letters in 1787. See Leopold von Schlözer,
Dorothea von Schlözer (Berlin: Deutsche Verlag, 1925).

[31]See Weierstrass' letters to his Göttingen colleague Lazarus Fuchs, in
"Briefe von K. Weierstrass an L. Fuchs," *Acta Mathematica*, 39 (1923), pp. 246–
56.

sufficient *aplomb* to undergo an *examen rigorosum*, and I fear that the unusual position, and having to answer, face to face, men with whom I am altogether unacquainted, would confuse me, although I know the examiners would do all they could for me. In addition to this, I speak German very badly . . .[32]

The Dean of the Göttingen Philosophical Faculty eventually gave in. In August 1874 Sofia Kovalevskaia was awarded the degree of doctor of philosophy in mathematics, *summa cum laude*, without an oral examination.[33] She was the first woman in the world outside of Renaissance Italy to receive her doctorate in that field, and one of the first to receive a doctorate in any field.[34]

The years of apprenticeship were finally over. As Iulia Lermontova put it: "Learning was done. Life had begun."[35] Sofia and Vladimir prepared to return to Russia, full of ambitious plans for their personal and professional lives. They both expected to be appointed immediately to prestigious teaching posts in St. Petersburg. Also, they seem to have been thinking seriously of embarking on their "new life" together, as real husband and wife.

Vladimir's letters to his brother during the Kovalevskiis' last months abroad were full of what one of his biographers exasperatedly calls "air castles."[36] Unfortunately, both Sofia and Vladimir soon learned that the reality of life in Russia was far different from their rosy imaginings. The next nine years were to be a time of increasing disillusionment, frustration, and eventual tragedy.

[32]Quoted in Leffler, "Sonya Kovalevsky," p. 188.

[33]*VP 1974*, p. 372.

[34]For descriptions of other women to distinguish themselves in academic fields, see H. J. Mozans, *Woman in Science* (1913) (Cambridge, Mass.: MIT Press Reprint, 1974), p. 53ff. and *passim*; S. Nikitenko, "Zhenshchiny-professora Bolonskogo universiteta," *Russkaia mysl'*, No. 10 (1883), pp. 253–91, and No. 11 (1883), pp. 85–127.

[35]Lermontova, p. 379.

[36]Shtraikh, *Sestry*, pp. 184, 221.

seven

MATHEMATICS IN ABEYANCE

Sofia and Vladimir Kovalevskii left Western Europe in late summer of 1874.[1] They spent the first weeks of their return to Russia at Sofia's family estate, Palibino. It must have been strange for her to wander around her old home, remembering with what excitement and anticipation she had prepared for her wedding and "liberation" six years before. Much had happened since then. She had experienced war and revolution firsthand during the months of the Paris Commune, suffered from the uncertainties of her ambiguous relationship with her husband, succeeded in convincing European mathematicians of her seriousness and capabilities, and attained one of her most cherished dreams—a doctorate in mathematics from Göttingen.

Now was not the time, however, complacently to view past achievements, tempted as she must have been by the adulation of her family circle at Palibino. Nor was it the time to sink back into the torpor of Russian country life. Kovalevskaia often commented that her life was a constant battle between the German industry of her mother's family and the Russian sloth of her

---

[1]This was the summer of the Populists' famous "Go to the People" movement. (See Venturi, *Roots of Revolution*, pp. 468–506, for details.) It is possible that Kovalevskaia returned to Russia at this time with a vague intention of being part of the movement to the people, and then hesitated at the last minute. Certainly, according to the progressive journalist L. F. Panteleev, she knew many of those arrested that summer. (L. F. Panteleev, *Iz vospominaniia proshlogo*, p. 504.)

father's. "In Russia I'm just not up to studying science," she would say, "there's just not that type of atmosphere here."[2]

At Palibino, Sofia was feted and honored. The whole family gathered to welcome her home. Aniuta and her husband came from St. Petersburg, where Jaclard was trying to establish himself as a French teacher, and brought their infant son with them.[3] Sofia's brother Fedia, now a young man of eighteen whom Sofia had persuaded to study mathematics at St. Petersburg University, came back to Palibino for his holidays.

After a few weeks' stay at Palibino for recuperation and reunion with the scattered members of the Korvin-Krukovskii family, Sofia and Vladimir went to St. Petersburg. They wanted to introduce themselves into scientific circles in the capital, and obtain university or academy posts as soon as possible. In late September 1874, they took an apartment in a building shared by different branches of Sofia's family.

Iulia Lermontova soon returned to Russia as well, with a chemistry doctorate from Göttingen under her arm. Unlike Sofia, Iulia did go through with her oral examination for her degree, although Weierstrass had tried to get her excused also.[4] Now, she too wished to obtain scientific employment in St. Petersburg, and so she moved in with Vladimir and Sofia.

At first, their reception by the St. Petersburg scientific elite was all that Sofia, Vladimir, and Iulia could have desired. They were invited to a gathering held by the famous chemist Dmitri Mendeleev, where they encountered luminaries of the mathematical, biological/paleontological, and chemical sciences. Kovalevskaia met the mathematician P. L. Chebyshev, among others, and engaged in a lively discussion with him until past midnight.[5]

Elizaveta Litvinova recounts that at first Chebyshev was the only Russian mathematician to be friendly to Kovalevskaia. Elizaveta explains this in part by Kovalevskaia's identification

[2]Quoted in E. El', "Iz vremen," p. 60.

[3]For Aniuta and Victor Jaclard's adventures during the intervening year see Knizhnik-Vetrov, *Russkie deiatel'nitsy*, pp. 203–04 and *passim*.

[4]*Briefe/Pis'ma*, pp. 45–6. There is a charming description of Lermontova's exam in her "Vospominaniia," *VIP 1951*, pp. 378–79.

[5]Late 1874 letter from Vladimir to Aleksander, Shtraikh, *Sestry*, p. 238.

with "the German tendency in mathematics," to which the Russians, even including Chebyshev, were at first hostile.[6]

By the "German tendency" Litvinova meant the school of "mathematical analysis for analysis' sake" of which Weierstrass was a leader. Chebyshev and the Russian school preferred a more concrete, down-to-earth approach, with a grounding in practical problems. Moreover they, like Vladimir's colleagues, had certain anti-German tendencies that were as much political and social as they were scientific. Consequently, many Russian mathematicians looked with suspicion on Kovalevskaia's German training.

The three friends were pleased by the reception at Mendeleev's. But to their surprise and chagrin, this cordiality was not followed by the brilliant job offers they had expected. Of the three of them, only Iulia was able to obtain a position even slightly commensurate with her training, and that only after years of trying. She worked in academician A. M. Butlerov's laboratory in St. Petersburg, an experience which she called "a true delight." In addition, she wrote for the review section of the *Journal of the Russian Chemical Society* in Moscow.[7]

Kovalevskaia, on the other hand, soon learned that any position in the Russian system of higher education was closed to her as a woman. This was typical of the situation encountered by educated women at this time. In order to teach in the higher grades of the *gymnasium* or in the university, one had to possess a Russian master's degree.[8]

Sofia was ready to take the examination, and had work which she could easily present as a master's thesis. But she discovered that women were legally forbidden to sit for the magistral exam. The only opening for her would be in the lowest grades of the women's *gymnasium*. "Unfortunately," Kovalevskaia remarked sarcastically, "I am not strong in the multiplication tables."[9]

[6]Litvinova, p. 43.

[7]Musabekov, p. 39ff.; Lermontova, p. 380.

[8]Exceptions to the Russian master's exam requirement were sometimes made for foreigners, but never for Russians, even if they had received their doctorates abroad.

[9]Quoted in Vorontsova, p. 147.

Vladimir was finding his prospects equally disappointing, though not for the same reasons as his wife. Russian geologists and paleontologists were resentful of his foreign degree and his boastful claims of superior expertise. After his disgrace at the Odessa examination, Kovalevskii's general reputation in their eyes was not good. In addition, Vladimir had left St. Petersburg under a financial cloud, and still suffered under the stigma of unpaid debts, including undischarged obligations to some of his former translators. Although he was awarded a prize by the Petersburg Mineralogical Society in December 1874, and finally received his magistral degree there in March 1875, he was not offered any position that he felt inclined to accept. In fact, he lamented to his brother that he regretted returning to Russia.[10]

In the first shock of their disappointment, Sofia and Vladimir made a decision which would significantly affect the course of both of their lives. It was decided that since neither of them had substantial private means, and since there were no immediate job prospects, they should devote their common energies toward accumulating enough capital through financial speculations to free themselves from the necessity of working at all. In what seems to have been a mood of almost manic optimism and euphoria, the couple determined that two or three years would be sufficient for them to attain financial independence.

It is difficult to determine which of the Kovalevskiis started their ruinous speculative dealings. The evidence seems to indicate that Vladimir had the original idea. The fact that in spite of her financial problems Kovalevskaia showed no interest in get-rich-quick schemes after she stopped living with Vladimir, while he was involved in them for the rest of his life, argues strongly against the theory that Sofia was the more eager of the couple.[11]

[10]Davitashvili, p. 322ff.; 28 September 1874 letter in *N N*, pp. 336–37.

[11]Two of Kovalevskii's biographers, Borisiak and Davitashvili, blame Kovalevskaia for what they claim was her instigation of the pair's "speculative trend." Borisiak says: "There can be no doubt that the soul of this 'new direction' was S. V., and V. O. was only her instrument, from love of her . . . S. V. succeeded in surviving the speculation crisis and returned again to her beloved science; V. O., as a man, was more closely involved with the 'new life', and died." (Borisiak, p. 42.) Davitashvili echoes this comment on pp.

Sofia, however, was a willing participant in the initial speculative schemes, and so cannot be exonerated from blame in the ensuing fiasco. In any case, whoever came up with the idea was able to infect the other partner with the mania. Sofia and Vladimir enthusiastically turned to financial speculation.

Thus began a new stage in the life of the Kovalevskiis. Mathematics and paleontology were for the most part put aside, the glittering salon world of the capital was eagerly courted, and Sofia and her husband entered a period of thinking of everything else but science. For Kovalevskaia, this time would extend for over five years; for Kovalevskii, it was to last the rest of his life.

This hiatus in Sofia's mathematical career is the subject of much discussion by those who write about her. Comments range from Eric Temple Bell's frivolous condemnation, to Beatrice Stillman's psychological theorizing, to P. Ia. Polubarinova-Kochina's enumeration of practical difficulties.[12] The person who was closest to Kovalevskaia in 1874–75, Iulia Lermontova, felt that Sofia's inactivity could be attributed largely to exhaustion:

> . . . as a whole our life in Berlin—in a bad apartment, with awful food, breathing unhealthy air, with uninterrupted and very exhausting work and in the absence of any kind of entertainment whatsoever—was without joy to such an extent that I looked back on our first months in Heidelberg as on a lost paradise. And Sofa was the same. When she received her doctorate in the fall of 1874, she felt such a loss of both mental and physical strength that for a long time after her return to Russia she couldn't undertake any work at all.[13]

The last two years in Berlin Sofia had been ill frequently, and she continued to ail in St. Petersburg. Her former adviser, Karl

326 and 331 of his biography. Semen Reznik, on the other hand, considers that Vladimir was without doubt the instigator. (*Vladimir Kovalevskii*, pp. 251, 263, 403.)

[12]Bell, *Men of Mathematics*, pp. 426–27; Stillman's Introduction to Kovalevskaya, *Russian Childhood*, p. 21 and *passim*; Polubarinova-Kochina, *Zhizn'*, p. 15.

[13]Lermontova, p. 387.

Weierstrass, was worried about her health. He wrote that he expected her to want to rest for a while, and strongly encouraged her to do so.[14] Of course, at the time he did not consider that her vacation from mathematics would be so prolonged, nor so complete as it turned out to be. Obviously, there was something deeper here than mere exhaustion and the inability to find work commensurate with her training, although these factors certainly must not be discounted.

One of Kovalevskaia's novellas, *A Nihilist Girl*, contains many autobiographical elements. Among them is what seems to be Sofia's own explanation for not having devoted herself to mathematics upon her return to Russia:

> After five years of solitary, almost reclusive life in a small university town, life in St. Petersburg immediately enveloped me and, so to speak, intoxicated me. Forgetting for the time [my mathematical studies], I now plunged into new interests, made acquaintances left and right, and tried to insinuate myself into the most widely varied circles. With avid curiosity I examined all manifestations of that bustling 'hurly-burly'—so empty in essence yet so attractive at first glance—which is called Petersburg life.[15]

This explanation seems to contain more than a grain of truth. Sofia was indeed "intoxicated" with life in the capital. She had always loved the theater, literary circles, and social interactions, but during her years abroad she had suppressed this longing for gaiety. Back in St. Petersburg, it resurfaced. She took on the trappings of a "society lady," supported her husband's plans to accumulate capital, and to a large extent lost touch with the scientific world.

At first, Kovalevskaia tried to combine mathematics with her salon life in the capital. She kept up her mathematical correspondence with Weierstrass, and talked of visiting him in the summer of 1875. She attended some meetings of Petersburg

---

[14]See Weierstrass' 21 September and 16 December 1874 letters to Kovalevskaia in *Briefe/Pis'ma*, pp. 48–52.

[15]*Nigilistka, VP 1974*, p. 90.

scientific societies, and continued to debate with Chebyshev occasionally. Moreover, she tried to find women students to tutor privately in mathematics.[16]

In September 1875, however, Sofia's father died. There had always been a strong bond of sympathy between them, and she and Vasilii Vasilevich had become even closer since she returned from abroad. His death was a considerable shock to her. In her grief, Sofia turned away from mathematics. She became engrossed in her new life, and from October 1875 to August 1878, Weierstrass heard nothing from his former pupil.

People who knew her during her immersion in the social world and the successful first stage of the Kovalevskiis' capitalist adventure, found Sofia much changed from the serious, singleminded student she had been in Berlin. Almost overnight she molded herself into a typical society lady, with most of the frivolous concerns common to that breed. She began to dress elegantly, though as always, untidily. She had weekly "at homes," and was seen regularly at the opera and theater.

This transformation from overworked student to seemingly carefree socialite was accompanied by an analogous change in the Kovalevskii marriage. Sometime in early 1875, Sofia and Vladimir decided to consummate their seven-year-old union, and live together as a normal pair.

The reasons for the decision are not clear. Possibly, neither of them made a conscious choice, but rather just drifted into a sexual relationship through inertia. Divorce was difficult, both were tired of living alone, and neither of them was in love with anyone else. In addition, Kovalevskaia seems to have wanted a child, although Vladimir did not.[17]

Whatever the reasons for the change in Sofia's life, they appear to have been accompanied by an alteration in personality as well. Sofia's friends found that her behavior was now more "feminine" and dependent. They felt that these characteristics were consciously assumed, and that Kovalevskaia deliberately acted out such a role in order to make the marriage work. Kovalevskii, in his turn, encouraged this role-playing in order

[16]*Briefe/Pis'ma*, pp. 70–1; Lamanskii, TsGALI, f. 280, ed. khr. 340.

[17]Vladimir to Aleksander, June 1875 letter in *N N*, p. 343.

to bind Sofia more closely to him, and make her feel that without him she would be helpless.[18]

Meanwhile, Vladimir was turning his hand to financial endeavors. He planned to resurrect his heavily indebted and virtually defunct publishing firm, and perhaps expand it as well. He wrote to his brother, outlining complex schemes which required more capital than the sum of his accumulated debts.[19]

Aleksander castigated his brother for what he called the couple's "speculative tendency." In her new role as feminine support and prop of her husband, Sofia answered the charge. In a letter which is pathetic considered in light of later events, Kovalevskaia referred to her "mathematical" solution for their lack of means, saying "the figures speak quite clearly, one only needs patience."

Sofia defended Vladimir's previous financial failures, and even went so far as to take the blame upon herself for the state of his publishing business in 1868–69. "Probably, he would have achieved excellent results if only he hadn't abandoned the business to the whim of fate at the worst possible moment," she wrote to Alexander. "He got no support from anyone, and I was a foolish fledgling without a thought of practical affairs."[20]

When General Korvin-Krukovskii died in September 1875, he left Sofia an inheritance of 30,000 rubles. It was to have been 50,000, but the general had already given Kovalevskii an advance of 20,000 rubles to pay off some of his old publishing debts. The inheritance was not large, but it would have been enough to live on if the Kovalevskiis had invested it carefully.

Unfortunately, careful investment of money was not characteristic of Sofia and Vladimir. They became so enthusiastic about plans for getting rich that they invested not only Sofia's money, but that of their family and friends as well. Sofia persuaded her mother, brother, sister, and her uncle, Fedor Shubert, to invest in a grandiose real estate scheme. They planned to build and then rent out apartments and public baths

[18]Litvinova, pp. 44–6.

[19]*N N*, pp. 339–43.

[20]14 November 1875 letter in *VIP 1951*, pp. 243–44.

in the area near St. Petersburg University. Aleksander Kovalevskii resisted for a while, but finally even he contributed some money, as did Iulia Lermontova.[21]

The idea of building houses was inspired by several sources. One was the mathematician Chebyshev, who had made himself rich by buying up and then reselling mortgaged country properties. He was reputed to own an estate in every province in Russia. Sofia intended to go Chebyshev one better. "I'll have a house on every street," she told Litvinova half-jokingly.[22]

Another source of the real estate idea was Kovalevskii's brother Aleksander, who made the mistake of informing Vladimir of his own plan to buy one or two houses for investment.[23] There were also several acquaintances of Kovalevskii who had had surprising luck with similar schemes. This was enough. The Kovalevskiis were seized with "building fever."

Sofia and Vladimir were not the only people of the sixties who tried their hand at business and industry in the 1870s. As the Soviet historian S. Ia. Shtraikh points out, a large portion of the liberal intelligentsia "abandoned politics, took up the industrializing fever, and occupied themselves with speculation."[24] Some former nihilists rationalized their activities theoretically. They reasoned that they needed to help capitalism along so that Russia would progress toward a (peaceful) social revolution more rapidly.[25]

Sofia and Vladimir saw no need for rationalizations. In fact, it is not clear that they even saw any contradiction between their present mercantile behavior and their political idealism. Certainly they continued to see themselves as nihilists.

At first Sofia and Vladimir were to all appearances successful in their financial manipulations. They moved to a large new apartment, and then to a house with a garden and orchard. They even had their own cow, which in the city was a sign of

---

[21]*N N*, pp. 358–60.

[22]Litvinova, p. 46.

[23]A. O. Kovalevskii, 7 August 1875 letter to Vladimir, *N N*, pp. 344–45.

[24]Shtraikh, *Sestry*, p. 236.

[25]Polubarinova-Kochina, "Zhizn'," *Pamiati*, p. 24.

prosperity. Of course, they had financed all this splendor with mortgages and loans, but no one knew that at the time. Everyone felt that they must be wealthy, and since the Kovalevskiis confidently expected to be rich soon, they did nothing to dispel rumors of their affluence.[26]

As was mentioned above, Sofia entered enthusiastically into the salon life of St. Petersburg. She had always had literary interests. Through her mother's family (which contained several artists and writers) and her sister Aniuta's friendship with Dostoevskii, Sofia was able to take up acquaintance with many writers. Soon, Dostoevskii, Turgenev, Saltykov-Shchedrin, Nekrasov, L. F. Panteleev, N. N. Strakhov and others were frequenting the Kovalevskiis' home.

The couple's circle of friends broadened even more in early 1876 when Vladimir encountered a fellow student from his law institute days, the speculator-jurist V. I. Likhachev, friend of the satirical writer Saltykov-Shchedrin.[27] Likhachev introduced Kovalevskii to A. S. Suvorin, a talented journalist. After a short acquaintance, the pair proposed a joint newspaper venture to Vladimir. Under their ownership, *Novoe vremia* (New Times) was to provide a radical alternative to the docile pro-government *Golos* (Voice), thereby attracting the readership of the liberal public.

In the eyes of one of Kovalevskii's biographers, Suvorin and Likhachev saw immediately that the sometime paleontologist's publishing expertise, eagerness to get rich, and lack of business acumen could be turned to their advantage. Vladimir helped attract radical publicists to the paper, and gave the pair a large loan, which he could ill afford, to finance the press. He undertook much of the day-to-day supervision of the paper himself. He was *Novoe vremia*'s night editor, wrote unsigned lead articles, solicited contributions, and penned popular-scientific articles under his own name.[28]

[26]Litvinova, p. 46; Kovalevskaia's account notes from the 1870s, on scraps of paper in the IML.

[27]Likhachev's wife, E. O. Likhacheva, was interested in the women's movement, and wrote an excellent series of books on the history of women's education in Russia. She was an acquaintance of Kovalevskaia.

[28]Shtraikh, *Sestry*, pp. 246–47; V. O. Kovalevskii, 24 March 1876 letter to Jules Walles, TsGALI, f. 1348 (Kollektsia pisem), op. 4, ed. khr. 28.

Sofia applauded the move into newspaper publishing. It increased her circle of literary acquaintances, and gave her the chance to try her hand at writing as well. In the two years that Kovalevskii worked for *Novoe vremia*, Kovalevskaia wrote four articles on popular scientific themes, and several unsigned theater reviews.[29] She was still an amateur at this time, though, and her efforts were not always successful. Sofia's work appeared rarely, and some of her attempts were rejected rather cavalierly.[30]

In 1877, when Suvorin and Likachev had achieved their goal of wresting a substantial part of the liberal reading public away from *Golos* and other newspapers, they began to change the political tendency of *Novoe vremia*. Instead of hinting at the need for reform, as before, editorial articles increasingly supported reactionary government positions. Saltykov-Shchedrin, Nekrasov, Turgenev and others at first tried to argue with the paper's owners. But when they saw it was useless, they ceased publishing there. The newspaper's political quality deteriorated with the departure of more progressive writers. By the 1880s, Georg Brandes, a Danish liberal critic who visited Russia, would call the paper "entirely without principle . . . more read than respected."[31]

The Kovalevskiis left *Novoe vremia* in 1877, even though they had more at stake than other contributors. The new reactionary tendency of the paper was no more acceptable to them than it was to other progressive intellectuals, but Vladimir had invested close to 20,000 rubles in the operation. Unfortunately, but not surprisingly when one considers Kovalevskii's previous record on business transactions, he never recovered that money from Suvorin.[32]

---

[29]See the 10 November 1876 and 10 March, 5 May, and 21 July 1877 issues of *Novoe vremia*. See also the list of possible articles by Kovalevskaia in [Polubarinova] Kochina, *Sofia Vasilevna Kovalevskaia*, p. 291. Polubarinova-Kochina describes some of these articles in detail on pp. 214–17, but none of them are particularly memorable or interesting.

[30]L. F. Panteleev, "Pamiati S. V. Kovalevskoi," *Rech'*, 29 January 1916.

[31]Georg Brandes, *Impressions of Russia* (New York: Crowell, 1889), p. 137. Brandes was a friend of Kovalevskaia.

[32]Shtraikh, *Sem'ia*, p. 279. For an account of Suvorin's career, see Effie

Vladimir and Sofia do not seem to have regarded this loss of capital as important, although their insouciance is difficult to comprehend. They went on with their plans to build houses, made feeble attempts to put Kovalevskii's old publishing business back on its feet, and continued to entertain an ever-widening circle of friends and acquaintances.

One of their visitors was the young Swedish mathematician Gösta Mittag-Leffler, a student of Weierstrass. Mittag-Leffler had come from his and Sofia's common adviser to try to interest her in mathematics again. He did not succeed, but he came away with an excellent impression of the twenty-six-year-old Sofia's knowledge, intelligence, and charm:

> More than anything else in Petersburg what I found most interesting was getting to know Kovalevskaia. . . . As a woman, she is fascinating. She is beautiful and when she speaks, her face lights up with such an expression of feminine kindness and highest intelligence, that it is simply dazzling. Her manner is simple and natural, without the slightest trace of pedantry or pretension. She is in all respects a complete 'woman of the world'. As a scholar she is characterized by her unusual clarity and precision of expression. . . . I understand fully why Weierstrass considers her the most gifted of his students.[33]

Although most of Kovalevskaia's other friends had no means of judging her mathematical abilities, Mittag-Leffler's description of her charm and exceptional liveliness was one with which they all would have agreed. Not all considered her beautiful, but everyone felt that Sofia lit up a room just by entering it, and that no conversation was dull if she was a part of it. L. F. Panteleev recalled that "she had only to appear somewhere and immediately she brought in some sort of special liveliness, and became the central figure of a circle of several surrounding people."[34] Apparently, she never lost this ability. The Swedish

---

Ambler, *Russian Journalism and Politics 1861–1881 (The Career of Aleksei Suvorin)* (Detroit: Wayne State University Press, 1972).

[33]G. Mittag-Leffler, 10 February 1876 letter to the Swedish mathematician Malmsten, quoted in "Weierstrass et Sonja Kowalewsky," p. 172.

[34]Panteleev, "Pamiati S. V. Kovalevskoi," *Rech'*, 29 January 1916.

feminist Ellen Key later called Sofia "the Michaelangelo of conversation."[35]

Sofia had a salon of sorts. That is, the Kovalevskii apartment became a gathering place for intellectuals and society figures. But the life of the Kovalevskiis was so haphazard, even when their financial affairs seemed to be going smoothly, that one never knew whether one's host and hostess would be available, or whether there would be a chair free of books or papers on which to sit. Litvinova remembered friends of the couple teasing them because their apartment always looked as if they had not yet properly moved in. Sofia and Vladimir would never have a comfortable, elegant home, their friends said, because they were gypsies at heart.[36]

Kovalevskaia gave birth to a daughter on October 17, 1878, at the age of twenty-eight. Sofia had had a difficult pregnancy, which left her with a weakness of the heart that possibly contributed to her premature death. Moreover, her labor was prolonged and unusually painful; she ailed for several months afterwards.[37]

The daughter, who was also named Sofia, was nicknamed Fufa to distinguish her from her mother. She was Kovalevskaia's only child, and both her parents adored her. But the advent of Fufa made the chaos in Sofia's and Vladimir's home even worse than before.

Kovalevskaia was passionately devoted to Fufa. A nanny had been engaged almost six months before the birth, several rooms were assigned exclusively for Fufa's use, and the most modern methods of child-rearing were employed in the infant's care. But Sofia refused to believe that anyone besides herself could do anything correctly for her baby. She became upset whenever Nanny or the wet-nurse went near Fufa, because she feared infection.

---

[35]"Extracts from Ellen Key's Biography of the Duchess of Cajanello," translated by Isabel F. Hapgood, in *Sonya Kovalevsky. Her Recollections of Childhood*, p. 299.

[36]Litvinova, p. 46.

[37]Undated letter from V. O. to A. O. Kovalevskii, quoted in Shtraikh, *Sestry*, p. 258.

Sofia was also anxious about her baby's intellectual level. Like many people at the time, she subscribed to the belief that one's mental state during pregnancy would affect the mind of one's unborn child. In those first few months of watching Fufa grow, Sofia remarked to Litvinova: "Thank heavens I had not completely lost my strength in the study of mathematics; now, at least, my little girl will inherit fresh intellectual capabilities."[38]

For a short time, observers thought that Fufa would inherit millions of rubles as well as her mother's intelligence, but the Kovalevskiis' construction empire came crashing down only months after her birth. Vladimir had overextended himself, staking nonexistent profits from as yet unbuilt homes to obtain further loans to buy more land. In addition, he still had bad debts from his publishing business. When Vladimir's creditors began to call in their loans, it was discovered that the Kovalevskiis' real estate empire existed only in Sofia's and Vladimir's imaginations.[39]

The Kovalevskiis' relatives and more well-intentioned friends had warned them of the danger of trying to get rich quick on borrowed capital. Sofia's mother Elizaveta Fedorovna cautioned her daughter just before her death in February 1879, and the publicist L. F. Panteleev also took it upon himself to chastise Sofia and Vladimir.[40]

But neither of the Kovalevskiis had listened to the dire predictions. Both obviously felt that with their combined wits and Vladimir's supposed business acumen, their success was assured. In any case, by 1878 Vladimir was so infected with speculative fever that it is doubtful whether he could have made himself stop.

The dissolution of their financial hopes affected Vladimir and Sofia quite differently. Kovalevskii began more and more to exhibit his tendencies toward nervousness, alternating states of depression and euphoria, and inability to concentrate on any-

[38]Litvinova, p. 47.

[39]There is a packet of bills, mortgages, and angry letters from creditors dating from this period in the Kovalevskaia papers, IML.

[40]Litvinova, p. 47; L. F. Panteleev, "Pamiati S. V. Kovalevskoi," *Rech'*, 29 January 1916.

thing for long. His letters to his brother became more absent-minded, rambling, and sometimes ranting; so much so that it was increasingly the task of Sofia to write to her brother-in-law to spare Vladimir the effort.[41]

As if the real estate fiasco was not enough to disturb Vladimir's high-strung personality, someone sent him a copy of the February 1879 issue of *Obshchee delo* (The Common Cause), a revolutionary paper printed in Geneva by Kovalevskii's old friend V. A. Zaitsev and others. One of the articles, "Something about Spies" (probably written by Zaitsev himself), was a direct attack on Vladimir. It revived the spy rumors, accused him of puffing himself up as a close collaborator of Darwin, said Sofia had been tricked into marrying him, paired him with Suvorin in the plot to turn around the politics of *Novoe vremia*, and claimed the two only fell out when Kovalevskii tried to cheat Suvorin over the press.[42]

It is unclear why Zaitsev revived the rumors at this time. The Kovalevskiis were still friends with Vladimir's former fiancée, the revolutionary Maria Mikhaelis-Bogdanovich, and her family, and regularly visited numerous people with ties to revolutionary circles. Moreover, Sofia was friends with the Chernyshevskii family and their cousins, the Pypins. So the rumors must not have been widely believed.

Vladimir was upset and almost unhinged by the combination of the financial disaster and the renewal of the slanderous rumors. He began picking quarrels with his wife, would write nothing coherent to his brother, and withdrew further into himself. He desperately tried to consolidate his debts, borrowing money from one creditor to pay another. But he only succeeded in getting his family deeper into debt.

Sofia, on the other hand, took the threat of bankruptcy calmly, stoically, even, it appears, with a certain amount of relief. For three years she had sampled the frivolity of the Petersburg social world. At first, she had been absorbed in it, but gradually, she had become impatient with its superficiality. She then began to search for more serious occupation.

[41]See letters in *VIP 1951*, p. 249ff. and in *N N*, p. 361ff.

[42]*Obshchee delo*, No. 20–45 (February 1879), article reprinted in full in Shtraikh, *Sestry*, pp. 262–65.

Kovalevskaia tried her hand at literature and journalism, renewed her radical contacts, and met some of the younger generation of revolutionary populists. She attended the "Trial of the 193" in the winter of 1877–78, and later wrote sympathetically about the radical defendants in *A Nihilist Girl*. Using her friend Dostoevskii's acquaintance with influential government officials, she was even able to help a friend of hers meet an imprisoned revolutionary for a fictitious marriage.[43]

Kovalevskaia was involved in the funding drive for the Higher Women's Courses which opened in St. Petersburg in 1878, and in 1878–79 was elected to membership in the college's twelve-person fund-raising committee.[44] Together with the Courses' founder, Nadezhda Stasova, Sofia's old tutor A. N. Strannoliubskii, and others, she busily sought out furniture, books, funds, and teachers for the school.

In the back of Sofia's mind was the idea that she would be asked to teach mathematics there. Unfortunately, she soon discovered that the government had no intention of giving women professorships at the Courses. Although the Ministry of Education grudgingly permitted the institution of the Higher Courses, they at first effectively prevented women from becoming professors there by refusing to permit women to sit for the master's exams. Women could thus be laboratory or classroom supervisors, but were barred from the regular faculty.

Kovalevskaia was disappointed when she was passed over for a position. She told Litvinova that if she had been hired, she would have abandoned all society life without regret. "Let me

[43]Some of this is described in *Nigilistka*, *VP 1974*, pp. 90–156. In the novel, however, Kovalevskaia does not discuss her own role in bringing her "nihilist girl" and the revolutionary I. Ia. Pavlovskii together. For this, see her letters to Dostoevskii, *VIP 1951*, p. 247 and notes 183(1) through 202(1), pp. 482–84, and chapter fourteen below.

[44]Z. A. Evteeva, ed., *Vysshie zhenskie (Bestuzhevskie) kursy* (Moscow: Kniga, 1966), p. 136; S. N. Valk, ed., *Sankt Peterburgskie vysshie zhenskie (Bestuzhevskie) kursy* (Leningrad: Izdatel'stvo Leningradskogo Universiteta, 1973), pp. 10, 174. Kovalevskaia was listed as an honorary member of the fundraising committee until 1918, when the Women's Higher Courses merged with Petrograd University (i.e., 27 years after her death). The Courses were sometimes called the Bestuzhev courses, after one of the first director/teachers, and the women who attended the Courses were sometimes colloquially called the "Bestuzhevki." For more information on the Courses, see P. N. Ariian, *Pervyi zhenskii kalendar'*, any year from 1899 to 1912, Directory of Women's Education.

Elizaveta Korvin-
Krukovskaia (Shubert),
Kovalevskaia's mother

General Vasilii Korvin-
Krukovskii, Kovalevskaia's
father

Margarita Smith,
Kovalevskaia's governess

Iosif Malevich,
Kovalevskaia's tutor

Sofia at the age of fifteen
(1865)

Anna Korvin-Krukovskaia
Jaclard, Kovalevskaia's sister. Anna was a writer, and
participated in the Paris
Commune of 1871.

Sofia Kovalevskaia in her mid-twenties.

The mathematician
Karl Weierstrass
(1880s), Kovalevskaia's
scientific adviser and
lifelong friend.

The mathematician Gösta
Mittag-Leffler as a young
man. Mittag-Leffler was
largely responsible for ar-
ranging Kovalevskaia's professorship in Stockholm.

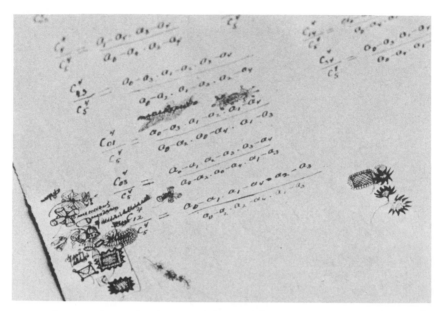

Typical page of Kovalevskaia's mathematical computations, complete with doodles. (Institut Mittag-Leffler archives—Angela Wanglert photo)

Kovalevskaia's household account book (she kept track of every penny, since she was always in straightened financial circumstances) lying on top of a page of intricate computations. (Institut Mittag-Leffler—Angela Wanglert photo)

Kovalevskaia and her friend, the Swedish writer
Anna Carlotta Leffler (the Duchess of Cajanello)

Sonnet   [boon] [weltering]
          [nestling]

To a young lady about to navigate
one of the heterodhedral concerts in
Balliol College.

Fair maid! whose voice calls music from the skies
(To labouring man, the aven's best, best pleasant boon,
Whose stops are various as the ... cheap ... morn,
And soft as dew that falls from pitying skies —
Let not false fears, thy worth ... ...,
... thy flight; dispel ... eyes ... boon,
Falls one, the rose's ... breath weltering ...
Or nestling breeze that in cool hollow dies?
Thee whose star shines bright o'er ... take
And thee who beautifiest glad Isis' shore,
Grant, I one joint harmonious garland bind!
Then cont with sounds our senses captive take
She in dumb numbers with Promethean power
Strike the full chords of all an all conquering mind

J. ...

New College.
Nov 8. 1886.

Sonnet by the British algebraist J. J. Sylvester. According to the accompanying letter to Mittag-Leffler, Kovalevskaia is the Muse of the Heavens mentioned in lines 9–14. (Institut Mittag-Leffler archives)

only be touched by mathematics and I forget everything else in the world," she said.[45]

These activities—dabbling in journalism, attending the Trial of the 193, keeping up old radical friendships, fund-raising for the Higher Women's Courses—kept Kovalevskaia from sinking completely into what she now called "the soft slime of bourgeois existence."[46] Interesting as these projects were, however, it became increasingly clear that they were not enough to satisfy her intelligence. During her pregnancy and the illness which followed it, when she was obliged to curtail her societal activities, Sofia started to evaluate her behavior. She became appalled at her relative mental stagnation during the past three years.

Sofia decided that she had to regain control of her life. She planned a return to mathematics and to the intellectual world in general. But her return would be slow and painful, and she would encounter many setbacks before finally becoming a full-time, professional mathematician.

[45]Litvinova, p. 48.

[46]Quoted in Vorontsova, p. 156. The translation is Beatrice Stillman's, in Introduction to Kovalevskaya, *Russian Childhood*, p. 23.

eight

A SLOW REAWAKENING

In August 1878, Kovalevskaia resumed her correspondence with her old adviser, Karl Weierstrass. Understandably enough, he was distant at first. He could not comprehend her three years of silence. But he was a kind man, and highly valued Kovalevskaia's mathematical talent. After some recriminations, he took up writing to her as if there had been no interruption.[1]

Weierstrass suggested that since Sofia had been away from mathematics for so long, she should work on some concrete problem to acquaint herself with recent research in her field. Weierstrass proposed a study of refraction of light in a crystalline medium. Sofia followed this suggestion, and within a few years produced "a sort of fourth dissertation, not completely independent of Weierstrass, though the new results obtained were entirely her own."[2]

Slowly, hesitantly, Kovalevskaia was returning to mathematics. But at first it was difficult for her to work, isolated as she was from other specialists with whom she could exchange ideas. Sofia needed the support of a mathematical atmosphere in which to work. The great Russian mathematician P. L. Chebyshev was able to assist her in this area.

When Kovalevskaia started to think about mathematics

---

[1]*Briefe/Pis'ma*, p. 80ff. The correspondence became frequent only in late 1880.

[2]Private communication from Roger Cooke, historian of mathematics and mathematician at the University of Vermont, 5 September 1981 letter.

143

again, she went to Chebyshev for advice. He suggested that she prepare a talk for the Sixth Congress of Natural Scientists, to take place in St. Petersburg at the end of 1879. She was nervous, but she unearthed her notebooks with her unpublished results from 1874, translated her dissertation on abelian integrals into Russian, and presented the work at the mathematical section of the Congress.[3]

The paper was well received. According to Litvinova, who was present, Kovalevskaia's ideas "were distinguished by their novelty and freshness," in spite of having lain in a dusty trunk for over five years. Mittag-Leffler, who also was there, went away more convinced than ever of Sofia's capabilities. He determined, if at all possible, to find a position for her, even at the risk of jeopardizing his own future. Gösta realized that his attempts to obtain a university job for a woman would make him unpopular in certain quarters, but he decided to proceed anyway.

Kovalevskaia was encouraged by the cordial reception of her talk and her mathematical correspondence with Weierstrass and now Mittag-Leffler, and Mittag-Leffler's assurances of support in her search for a post. In light of her colleagues' enthusiasm, Sofia "was able to witness calmly the public sale of the last of her possessions several weeks later. She said that she would soon completely forget that she had been the sometime happy owner of [several] stone houses."[4]

The sale of the Kovalevskiis' possessions at public auction was the final blow to their fortunes. It also seems to have nudged Vladimir psychologically over the edge. He became even more unstable, as his letters to his brother during this period show. He was withdrawn, increasingly unable to concentrate, and had periodic fits of alternating euphoria and depression.[5]

Sofia was worried about Vladimir's inability to face the real-

[3]See A. V. Pogozhev, *Dvadtsatipiatiletie estestvenno-nauchnykh s"ezdov v Rossii, 1861–1886* (Moscow: Frish, 1887), pp. 208–09.

[4]Litvinova, p. 49; G. Mittag-Leffler, "Weierstrass et Sonja Kowalewsky," p. 173.

[5]See letters in *N N*, p. 365ff.; and Sofia's letters to A. O. Kovalevskii, *VIP 1951*, pp. 249–55.

ity of their bankruptcy. She tried to turn Vladimir's thoughts toward a return to paleontology, but he was unresponsive. Sofia consulted a specialist in nervous disorders, who had been recommended to her by Sechenov. "It is extremely important to fortify Volodia's [Vladimir's] nerves somehow," she wrote to her brother-in-law.[6]

The Kovalevskiis moved to Moscow to escape the scene of their financial disaster. Vladimir, Sofia, and her brother Fedia took out a 65,480 ruble loan to consolidate all the Kovalevskii debts, and pay off their St. Petersburg creditors. Fedia guaranteed the loan, because otherwise no one would have lent money to a couple with the liabilities of Sofia and Vladimir.[7]

The Kovalevskiis planned to take up housekeeping on a much smaller scale in Moscow than they had in St. Petersburg. At the same time, they hoped Vladimir would obtain some sort of job in the academic community. Iulia Lermontova moved with them to help out with both financial and moral support. She rejected an offer of a job as a laboratory instructor in the Higher Women's Courses in order to do so.

Iulia's decision to leave St. Petersburg led to furious attempts by the famous Russian chemist A. M. Butlerov to get Iulia to reconsider. When he did not succeed, he wrote his Moscow colleague V. V. Markovnikov to complain. The latter replied: "Here the whole reason lies in Sofochka Kovalevskaia. If it were not for her, Lermontova would be in St. Petersburg. That madame, making use of Iulia Vsevolodovna's kindness, systematically exploits her."[8]

---

[6]16 January 1880 letter from Kovalevskaia to Aleksander, *VIP 1951*, p. 252.

[7]The loan paper, signed by Fedor Korvin-Krukovskii, Sofia and Vladimir, is in the Kovalevskaia papers of the IML. The three apparently asked for such an odd amount because Sofia and Vladimir had no money to pay the first year's interest in advance—i.e., 9% of 60,000 rubles, or 5,400 rubles plus transaction fees.

[8]Quoted in Musabekov, p. 45. Musabekov justly comments: "Here Markovnikov is not completely respectful to Kovalevskaia [Musabekov refers to the derogatory use of the second-level diminutive 'Sofochka' especially in contrast to the respectful use of 'Iulia Vsevolodovna' for Lermontova], but one can partly justify him. . . . Sofia Vasilevna sometimes abused [Iulia's devotion]" (p. 45).

Perhaps it is true that Kovalevskaia occasionally took advantage of Iulia's good nature. But Iulia does not seem to have minded. She was Fufa's godmother (the godfather was I. M. Sechenov), was grateful to Sofia for starting her on her university studies, and was fond of Vladimir. Moreover, she claims not to have wanted a boring laboratory job supervising beginning chemistry students in St. Petersburg. Although she worked for a time in Markovnikov's Moscow laboratory, his projects did not interest her. Lermontova soon left chemistry altogether to take up agriculture and cheese-making at her estate outside Moscow.[9]

Meanwhile, as if the failure of his first publishing business, his bad experience with Suvorin, and the collapse of his real estate fantasies were not enough for him, Vladimir started one last attempt to make a fortune. While his brother Aleksander was maneuvering to get Vladimir a position at Moscow University, P. I. Bokov (the first husband of Maria Bokova-Sechenova) introduced Vladimir to the oil entrepreneurs Viktor and Leonid Ragozin.

The Ragozins' sentiments on the occasion are unknown. There is some evidence that they were clever and devious enough to spot a likely dupe when they saw one. They observed Vladimir's mental instability and feverish desire to get rich. They seem to have felt that both these facets of Kovalevskii's personality could be turned to their advantage, and promptly invited him to take a directorship in their company.[10]

Vladimir seems to have had enough sense left actually to debate with himself about whether he should accept the position or wait for Aleksander to arrange a university post. He was even slightly suspicious at first. Viktor Ragozin could not be pinned down on exactly what his job would be, and was promising inordinately large stock dividends—sixty to seventy percent. But in the end Kovalevskii's desire to make up for his previous failures and his mania for commercial transactions

[9]Lermontova, p. 381.

[10]Shtraikh, *Sestry*, p. 269ff.; Reznik, *Vladimir Kovalevskii*, p. 277ff.; Vorontsova, p. 160ff.; Leffler, "Sonya Kovalevsky," p. 195ff. Reznik is the only commentator who feels that the Ragozins were honest, and that Vladimir was imagining deceit where none existed.

prevailed over his limited common sense. He joined the Rago-zin company in May 1880.[11]

Sofia was unenthusiastic, especially at first, about her hus-band's new venture. She mistrusted Viktor Ragozin on sight, and was suspicious of his friendliness to her poor husband.[12] Sofia tried to focus Vladimir's energies solely on the university, but was unsuccessful.

Vladimir's nervous state made him reluctant to share his thoughts with anyone. He was ashamed of the mess he had made of his family's finances, and so did not want to talk to his relations and friends. This led to further misunderstandings. Sofia was insulted by his taciturn behavior, and her personal affront blinded her to the fact that Vladimir was no longer entirely responsible for his actions.[13]

Sofia's failure to comprehend Vladimir's mental state was not only attributable to personal dudgeon. At the time, Kovalevskaia was not devoting all her attention to Vladimir's problems. As she had told Litvinova, once mathematics touched her again, she tended completely to neglect every-thing else.

Sofia had come to Moscow with the full intention of partici-pating in scientific life. She attended meetings of the Mathe-matical Society and gave papers there, although she was not elected to membership until March 29, 1881. She had lengthy discussions with Moscow mathematicians, and even per-suaded some of them to look favorably on her former teacher Weierstrass' method of analysis.[14]

Kovalevskaia determined to persuade the university to let her take the master's examinations.[15] She obtained the willing support of several mathematicians of the Moscow faculty—

[11]See letters in Shtraikh, *Sestry*, p. 269ff.; *N N*, p. 369ff.

[12]Leffler, "Sonya Kovalevsky," pp. 195–96.

[13]*Idem*; Litvinova, pp. 49–52.

[14]A. G. Stoletov, "Russkaia zhenshchina-matematik S. V. Kovalevskaia," *Nauka i zhizn'*, No. 23, 8 July 1891, p. 355; D. A. Tarasov, "Professor S. V. Kovalevskaia," *Russkie vedomosti*, 30 January 1901.

[15]Marie Zebrikoff, "Russia," in Stanton, *The Woman Question in Europe*, p. 417. Litvinova was also preparing herself for the exam at this time, but neither of the two women was ever permitted to take it.

N. V. Bugaev and A. Iu. Davidov, in particular—and made an official petition to the Minister of Education asking for permission to take the exams. She withdrew it almost at once, though, because the embryologist A. I. Babukhin and the gynecologist V. F. Snegirev told Vladimir that her application would jeopardize his chances at the university.[16]

Kovalevskaia was impatient with the delay, but resigned herself to waiting. A few weeks later, when Vladimir's appointment to the university had become "extremely likely," she tried again. She had the support of the Moscow mathematics department. Professor Davidov and even the university rector N. S. Tikhonravov appealed personally to the Minister, but to no avail. She was told that the Minister, A. A. Saburov, had said that "both I and my daughter will have a chance to get old before women will be admitted to the university."[17]

Kovalevskaia poured out her bitterness at her frustrated plans in a letter to Vladimir's brother:

> It is doubly annoying for me, because it will never be so right for me to take the examinations as it would be now. This summer [1880] I succeeded in preparing myself so well, that I could appear for the exam this very moment, especially since the mathematicians here all relate to me sympathetically, and would make only fair demands [i.e., would not increase the exam's difficulty, or give obscure questions on an unfamiliar topic so that she would fail]. Besides, they acknowledge that my (not yet published) work, an extract of which I gave this past winter at the Congress, would be fully suitable as a master's dissertation.[18]

Kovalevskaia decided that her only recourse was to prepare as many mathematical works as possible, "in order to keep up our women's reputation."[19] To accomplish this, she needed to

[16]Kovalevskaia, letter to Aleksander, *VIP 1951*, p. 253. This and the two following letter citations have been tentatively dated October 1880 by Shtraikh.

[17]*Ibid.*, p. 254.

[18]*Ibid.*, pp. 254–55. The reference is to Kovalevskaia's work on abelian integrals.

[19]*Ibid.*, p. 245.

reestablish contact with the European mathematical community. She wrote to Weierstrass, asking him whether she could visit him in Berlin that fall.

Weierstrass replied in the negative, saying that he would be too busy, but his letter reached Moscow too late to stop Kovalevskaia.[20] With characteristic impetuosity, Sofia had already left for Berlin. Without waiting for a reply from him to see if her visit would be convenient, she swept off at the end of October 1880, leaving two-year-old Fufa with Iulia Lermontova for over two months.

Even if she had received Weierstrass' letter in time, Sofia probably would have gone to Berlin anyway. She had been away from mathematical research for too long. She needed to see Weierstrass in person, in order to convince him that this time she was finally serious, and intended to return to mathematics permanently.[21]

Once Kovalevskaia was actually in Berlin, Weierstrass was happy to see her, although he could not devote as much time to her as he wished because of his busy schedule. Sofia's visit, though ill-timed, was valuable for him as well as for her. She resumed her role as the sounding board for his ideas, and amazed him by her unique approach to their mutual interests. He was able to refer her to work closely related to her study of refraction of light in a crystalline medium which had been published or circulated in manuscript form during the time she had been away from mathematics.[22] Sofia returned to Moscow mathematically refreshed, with her self-confidence restored.

This interlude in Berlin made Kovalevskaia better able to cope with the increasing tensions of her family life. When Vladimir and Sofia moved to Moscow, the two of them had tried to stay together for Fufa's sake, but the efforts were feeble and misunderstandings abounded. Although the Kovalevskiis still lived together (along with Lermontova and her sister), the marriage was fast coming to a close.

[20]Weierstrass to Kovalevskaia, 28 October 1880, *Briefe/Pis'ma*, pp. 82–3.

[21]Mittag-Leffler, "Weierstrass et Sonja Kowalewsky," p. 175.

[22]Weierstrass to Kovalevskaia, 31 October 1880 and 1 February 1881 letters in *Briefe/Pis'ma*, pp. 85–7.

In a last bid to save her marriage, Sofia put aside her scruples about the Ragozins and attempted to interest herself in inventions for the company. She succeeded for a while, by concentrating on the applied mathematics aspect, and even invented some sort of illumination device.[23] Lermontova also became fascinated with inventions. She had some dealings with the Ragozin company through her laboratory, and started working on more efficient methods of petroleum distillation.[24]

Elizaveta Litvinova describes the Kovalevskii-Lermontova household at this time (mid to late 1880) as hectic, but optimistic. The menage lived precariously on Vladimir's earnings from the Ragozin position and the Lermontovas' small inheritance. Sofia, Vladimir, and Iulia would work on their inventions all day, and in the evening dream of how their circumstances would change if one of their devices succeeded.

Iulia's sister Sonechka "represented the conservative element;" she would mock their fantasies goodnaturedly, and keep the baby occupied. Tiny Fufa also participated: "The little girl had hardly started to talk but, imitating the adults—father, mother, godmother—she took a scrap of paper and a pencil, drew some squiggles and said: *appaliat* [the baby version of *apparat* = apparatus]; obviously she often had occasion to hear the word."[25]

The pleasant state of affairs decribed by Litvinova did not last long. Vladimir, in an effort "to combine service to geology with service to Mammon," as Sofia put it in an October 1880 letter to her brother-in-law, had madly bought up shares in the Ragozin company. While he waited for confirmation of his university appointment, he purchased at least twenty-five shares costing over 20,000 rubles.

Since Vladimir possessed nothing like this sum, he borrowed money, mortgaged the shares he had bought to buy more, and also committed the nonexistent sixty percent profit of his mortgaged shares to further share purchase.[26] He should not be

[23]There are sketches of such a device in the Kovalevskaia papers, IML.

[24]Litvinova, pp. 51–2; Lermontova, p. 381.

[25]Litvinova, p. 51.

[26]Letter from Vladimir to Aleksander in Shtraikh, *Sestry*, pp. 269–70.

judged too harshly, however. What in a normal person could be rightly termed dishonest was in Vladimir simply a further proof of his deteriorating mental condition. He truly does not seem to have understood that what he was doing was unethical, if not downright illegal.

In view of their previous experience with Vladimir's enthusiasms, one would think Sofia, Iulia and Aleksander would have been more wary of any project in which he engaged. To be sure, Sofia and Aleksander repeatedly tried to get him to desist, especially after his appointment to the Moscow University geology faculty was approved in late December 1880. They did not make any firm demands, however, or give him any ultimatum, and Aleksander and possibly Iulia even seem to have loaned him more money.[27]

Kovalevskii's behavior was more erratic than ever. His position at the university was supposed to begin in January 1881, but he delayed his return from a trip abroad for the Ragozins.[28] He seemed to have fallen into a mood of total apathy, making no payments on his loans, and not even informing the university of his delay.

When Kovalevskaia returned from Berlin, she had to deal with angry creditors and a university faculty indignant at Vladimir's absence. She handled the matters with apparent calmness, but complained to her brother-in-law about her husband's strange attitude:

> Vladimir Onufrievich relates to all this with such outrageous flippancy, that he answers my letters and telegrams only a week and a half after he receives them . . .
>
> The university, it stands to reason, is also not praising him. In a word, this is all very depressing, and if last year's experience [their Petersburg bankruptcy] had no effect on him, then really there's nothing else we can expect [from him].[29]

---

[27]See A. O.'s letter to his brother, end of 1880, in *ibid.*, p. 273.

[28]Shtraikh, *Sestry*, p. 274. On this trip Vladimir spent time with several people who had ties to Zaitsev's revolutionary circle. Obviously, Zaitsev's revival of the old spy rumors had had no effect on these relationships.

[29]Kovalevskaia, January 1881 letter to Aleksander Kovalevskii, *N N*, p. 374.

Then, as if Vladimir's absence, attitude, and the gathering creditors were not enough, Leonid Ragozin told Sofia that her husband owed the company 17,000 rubles!

Sofia was completely exasperated. When Vladimir returned from his trip, they decided that they would try living apart. According to Litvinova, "it was still a tiff rather than a rupture," but it was soon made permanent.[30] Sofia and Vladimir corresponded, met abroad, and sometimes referred to one another affectionately in letters. But they had tacitly acknowledged that without love on either side, it was useless to try to patch together a marriage which should never have been consummated in the first place.

The Kovalevskii household dissolved in March 1881, around the time of the assassination of Tsar Alexander II.[31] Vladimir took Sofia and Fufa to the train station, where he left for Odessa to visit his brother, and mother and daughter went to Berlin. Kovalevskaia was on her own again, at the age of thirty-one. Although she was unhappy (Anna Carlotta Leffler says she cried half the way to Warsaw), she had the thought of unlimited immersion in mathematics to console her.[32]

As she moved away from Moscow, Kovalevskaia must have felt regret and bitterness over her wasted years. In fact, she once told Litvinova that the only good thing to come out of

[30]Litvinova, pp. 52–4.

[31]Nechkina puts forward the theory that the Kovalevskiis' abrupt departure from Moscow in March 1881 had something to do with the wave of reaction following the assassination of the tsar. ("Obshchestvennaia i literaturnaia deiatel'nost' S. V. Kovalevskoi," *Pamiati*, p. 92; repeated in *VP 1974*, pp. 501–02.) She points to the fact that, according to Litvinova, they left in a hurry—there was still food on the table, nothing was put away, "it seemed as if the owners of the apartment had just stepped out for a minute and would return" (Litvinova, p. 54). But she gives no further evidence, merely stating justly that Kovalevskaia's political commitment is rarely acknowledged adequately by historians. (Nechkina apparently missed a vague allusion by Sofia's German cousin which might have been used to support her argument. When Adelung saw Sofia in late 1880, looking so worried, she thought at first Sofia *was in some political trouble*, then discovered that the trouble was financial ("Jugenderinnerungen," p. 419).)

It is possible that Nechkina's theory may be true. Certainly, the Kovalevskiis' behavior at that time is puzzling. However, there does not seem to be enough evidence to judge.

[32]Leffler, "Sonya Kovalevsky," p. 198.

them was her daughter.[33] She must have thought with sadness of Vladimir, and with affection of the Lermontovas. But her strongest emotions were probably relief and anticipation. She had felt increasingly uncomfortable in the society lady mold, in spite of her social success. She now knew that such a life could never satisfy her; only a life of productive mathematical work could do that.

A life as a professional mathematician was not yet guaranteed to Kovalevskaia, but prospects were opening up. She was making progress on two problems she had made her special concern—refraction of light in a crystalline medium and revolution of a solid body about a fixed point. Weierstrass had found an inexpensive apartment for her in Berlin, and awaited her arrival with impatience. In addition, Weierstrass' former student, Gösta Mittag-Leffler, was attempting to arrange a position for Kovalevskaia.

Sofia had begun a correspondence with Mittag-Leffler in October 1880.[34] Since his visits to St. Petersburg in 1876 and especially 1879, Mittag-Leffler had been determined to help find Kovalevskaia a university level place as a professor of mathematics. Upon learning of Sofia's *de facto* separation from her husband, Gösta accelerated his efforts. He now was optimistic that he could obtain a post for her in Helsinki, Finland, where he was teaching.

Nothing was certain yet. The next two years would witness many setbacks and disappointments, and would culminate in the tragedy of Vladimir's suicide. It would be two and a half years before Kovalevskaia finally was offered a university post, longer still before she made significant progress on the rotation problem.

Yet her future direction was by this time clear. Mathematics was to be Sofia's life, and other concerns would have to take second place, at least until she had established herself professionally in the world of European mathematics.

[33]Litvinova, p. 50.

[34]Both sides of this correspondence—from Kovalevskaia to Mittag-Leffler and from Gösta to Sofia—are in the general letter collection of the IML. Excerpts have been published by Anna Carlotta Leffler in "Sonya Kovalevsky," and by Polubarinova-Kochina, "Iz perepiski S. V. Kovalevskoi," *Uspekhi matematicheskikh nauk*, 7, No. 4(50) (1952), pp. 103–25.

nine

A VAGABOND
MATHEMATICIAN

Sofia Kovalevskaia and her daughter Fufa arrived in Berlin in late March or early April 1881, and settled into the apartment Weierstrass had found for them. They brought with them a governess, so Sofia was able to devote all of her energies to mathematics. She saw Weierstrass frequently, studied various works on the light refraction problem, and kept up her experiments on electricity. Later, though, she fell back under Weierstrass' influence, and curtailed her practical research.[1]

Kovalevskaia continued corresponding with Mittag-Leffler on the possibility of a position at Helsinki University, although the prospects for successful conclusion of the negotiations were dwindling. The problem, it seems, was not with the mathematicians, who were convinced of Sofia's qualifications. Nor was the obstacle Sofia's sex—the Finns had a record of advanced views on the woman question.[2] Rather, the Finns feared that Sofia, as a Russian nihilist, would come armed with revolutionary ideas.

Finland was under Russian control at the time, but the yoke was comparatively light. Helsinki University, for example, had an autonomous position that was unique within the Russian empire. Mittag-Leffler wrote that the Finns feared to antagonize Russian officials by Sofia's appointment, especially since they felt that she would attract politically minded women

[1]Litvinova, p. 54.

[2]Helene Lange, *Higher Education of Women in Europe* (New York: Appleton, 1897), p. 114.

155

to her classes. Mittag-Leffler apologized to Sofia, and suggested that perhaps he could arrange something in Stockholm instead.[3]

Sofia was disappointed but, as she replied to Mittag-Leffler, she was not particularly surprised. Nor was she hopeful about Stockholm either:

> As to your excellent plans with regard to Helsinki . . . I never seriously believed in them in spite of very much wanting them to work out. I also do not intend to place too much hope on Stockholm; although I admit that I would be ecstatic if I were given the opportunity. . . . I would be delighted to open a new career for women . . . But I repeat, I do not want to abandon myself too much to [thoughts of] this wonderful project. It probably will have the same fate as the majority of wonderful projects in this world.[4]

The polite resignation expressed in this letter was not a reflection of Kovalevskaia's real feelings. She truly wanted a teaching job, and as the summer wore on, her eagerness won out. She wrote to Mittag-Leffler again, saying that she would take any position, no matter how lowly, as long as it was in an institution of higher education. She repeated her desire to open the university fully to women as students: "As it stands now this access is an exemption or a special kindness that can always be revoked, and this is what has happened at most German universities."

So great was Kovalevskaia's zeal that when Mittag-Leffler sounded her out on the possibility of assuming a post as *privatdocent*, which traditionally had no salary the first year, she immediately wrote him that she would accept such an offer.[5] There would be the fees paid by her individual students, and she reasoned that she could sell the jewelry her mother had left her to supplement her meager income.

Apparently, Kovalevskaia showed Weierstrass the draft of her letter to Mittag-Leffler. He became afraid that the Swedish

[3]Mittag-Leffler, 23 March 1881 letter to Kovalevskaia, Mittag-Leffler papers, IML.

[4]Kovalevskaia, 7 June 1881 letter to Mittag-Leffler, IML.

[5]Kovalevskaia, 8 July 1881 letter to Mittag-Leffler, IML.

mathematician would hurt his own position by his partisanship of Sofia, so she reluctantly added a warning that he not do too much too soon. Perhaps it would be better to wait until she finished her current projects, she wrote, because "if I succeed in working them out as well as I expect, then they will help achieve the desired goal [of silencing opposition]."[6]

Weierstrass' fears for Mittag-Leffler's position were not unfounded. There was much resentment against Gösta in Helsinki, because some people there thought that his place should have been awarded to an equally qualified Finn.[7] His involvement in a new controversy could perhaps further weaken his position.

There is also the possibility that Weierstrass discouraged the plans because of his old-fashioned notions of morality. He thought Sofia's place was with her husband. After meeting with Vladimir in mid-1882, however, Weierstrass decided that the couple had nothing in common, and that Kovalevskii would not support Sofia's desire to work in mathematics. The meeting, which took place in Berlin when Vladimir was on one of his trips abroad for the Ragozins, removed what might be considered Weierstrass' "moral" objection to Kovalevskaia taking a position. But he still was uneasy about Sofia living apart from Vladimir, and was afraid gossip would damage her reputation.[8]

Moreover, Weierstrass felt that since Sofia had been away from mathematics for so long, she needed to bring at least one of her current mathematical projects to completion. There would be enough opposition as it was on the basis of her sex, Weierstrass thought. It would be unwise to give her enemies the opportunity to say that Sofia had written nothing in over seven years.[9]

When Kovalevskaia offered to forego a salary temporarily, and conceded that it might be better to wait until she had done

[6]Kovalevskaia, 8 July 1881 letter to Mittag-Leffler, IML.

[7]Gustav Elfving, *The History of Mathematics in Finland, 1828–1918* (Helsinki: Societas Scientiarum Fennica, 1981), p. 72ff.

[8]Weierstrass, 14 June 1882 letter to Kovalevskaia, *Briefe/Pis'ma* pp. 93–4.

[9]Kovalevskaia, 21 November 1881 letter to Mittag-Leffler, IML.

more work, she was counting on receiving adequate funds from Kovalevskii. What she did not realize at the time was that Vladimir had gone so far into debt again that he had used up almost everything remaining to Sofia, including her mother's jewelry.[10]

At about the same time, she learned from Iulia Lermontova that Vladimir had been seriously ill that spring but had not written to inform her. This hurt Sofia far more than his clandestine disposal of her possessions. In spite of their differences, she had felt that they were at least good enough friends for Vladimir to have sent for her when he was ill.

Some of Kovalevskaia's letters during the summer of 1881 indicate that she might have lived with Kovalevskii again if only he had asked her to, and had given her assurances that he would not hinder her mathematical work. In one letter (undated) she suggests that he leave Russia and presumably come wandering with her in Europe: "You know you are keen on life abroad," she wrote.

Not only did Vladimir refuse this request, but he also hinted that she as a woman was wasting her time on mathematical studies. (Vladimir's earlier staunch support of women's rights seemed to vanish as he became more unstable.) Sofia acknowledged that no woman had yet made a name for herself in mathematics, but added that she hoped to be the first to do so:

> So far I have had three months (and those with notable hindrances) and truly, it seems to me that I have accomplished quite a lot, although I still don't have final results. But you know, when I do have them, they will immediately give me different status and a different name. Weierstrass says that my present work (in the event of full success) will be one of the most interesting results of the last decade.[11]

The tone of this letter alternates between sarcasm and tenderness. Sofia gives the impression that she regrets parting from Vladimir and will return to him, provided that he will make a commitment to family life and allow her to continue

[10]Vorontsova, p. 171.

[11]This and the preceding two letters are in Shtraikh, *Sestry*, pp. 276–79, *passim*.

with her work. She is willing to make compromises, if he will only meet her half-way.

By this time, however, there was no going back. Vladimir was confused and disoriented, and already did not know what he wanted. He took to answering Sofia's letters only after weeks' delay, and then his answers would be ambiguous. She proposed that she and Fufa go home to him in August. He delayed answering, sent them no money, and finally telegrammed what Sofia called "some kind of lofty tract on mutual freedom," telling her that she should do whatever she wanted.[12]

Kovalevskaia was furious, hopeful, tender, and exasperated by turns. She was not passionately in love with Vladimir, but she felt affection for him, and worried about his welfare. She kept up a chatty, determinedly cheerful (though occasionally angry) correspondence with him throughout 1881 and 1882. But she rarely received replies.[13]

In late autumn 1881, Kovalevskaia moved to Paris to work with mathematicians there and be near her sister Aniuta. Three-year-old Fufa and her nursemaid accompanied Sofia to Paris, where the three lived in a dreary set of furnished rooms. Although Aleksander Kovalevskii and Iulia Lermontova repeatedly offered to take Fufa, Sofia was reluctant to give her up. She was comforted by her daughter, and felt that the child needed her mother in these early years.[14]

At first, it was difficult for Sofia to concentrate on her mathematics. She was worried about Vladimir, and her own financial circumstances caused her anxiety as well. Moreover, she was reluctant to contact the French mathematicians either because she was shy, or because she correctly sensed that they would feel awkward with a woman on her own. She appears to have spent her first months in Paris isolated from her colleagues. Her circle consisted of Aniuta, Vladimir, whose business some-

[12]Letter in *ibid.*, p. 279.

[13]See Kovalevskaia's letters in Shtraikh, *Sestry*, p. 276ff.; *VIP 1951*, pp. 259–61; Polubarinova-Kochina, *Zhizn'*, p. 28ff.

[14]Kovalevskaia's diary entries for 1881 and early 1882, Kovalevskaia papers, IML.

times took him to the city, and their Paris friends in the Russian émigré community.

Kovalevskaia's diary at this time is full of worry about political events in Russia, Vladimir's mental state, and the Ragozin company. One of her new acquaintances turned out to be a tsarist spy, and this revelation caused her and her friends much anxiety. They had to be constantly on the alert for fear that their political conversations would be reported.[15]

Kovalevskaia and Iulia Lermontova tried to keep informed of Vladimir's activities with the Ragozins. At times, they were hopeful. They reasoned that if the Ragozins were keeping Vladimir on in spite of his tremendous debt to the company, they must feel that Vladimir was useful and would eventually be able to pay off his debt.

But Kovalevskaia's friends increasingly warned her about any association with the Ragozins. Herzen's son-in-law, the French historian Monod, told Sofia to get Vladimir out of the company as soon as possible. "He said horrible things about the [Ragozin] directors," Sofia noted in her diary on January 26, 1882.

Kovalevskaia tried to tell herself that the poor state of affairs of the company had been exaggerated. Nevertheless, she conveyed the warnings to Vladimir. Vladimir did not answer her letters, nor did he write to his brother. He seemed to be in an unnatural state of euphoria and feverish optimism, Iulia wrote from Moscow.

As if the worry about Vladimir and the political situation were not enough, Fufa became sick. She was seriously ill, and at first Sofia and the doctor feared cholera. The child recovered, but Sofia reluctantly decided to send her to Aleksander Kovalevskii in Odessa for convalescence. Aleksander had been sending money for Fufa, but he could not continue to do so, and Fufa needed more care than her mother could provide. On March 1, 1882 Kovalevskaia sent her daughter back to Russia with her nurse. The two did not live together again until several years later, after Sofia had settled in Stockholm.[16]

[15]15 January 1882 diary entry, IML. See also 22 May 1882 diary entry on political troubles.

[16]Diary entries for 28 and 30 January, and all of February 1882, IML. In later

Kovalevskaia was upset by Fufa's departure, and felt more lonely than she had ever felt in her life. She wrote a depressed letter to Clara Weierstrass, who was exasperated by the self-pity and gloom of her beloved "Sonia." In an effort to scold her into a more optimistic frame of mind, Clara wrote: "What are you writing such a melancholy letter for? Pack up your papers and come to Berlin. You can live like a hermit here, too, you know."[17]

But Sofia was in no mood to be cheered by Clara's gentle teasing. She had begun to suspect that Vladimir was having an affair with an Englishwoman he had met on his travels for the Ragozin company. The Ragozins encouraged her suspicions, possibly so that she would be thrown off-balance and not look further into the shady dealings of the oil company directorate. In any case, at this point Sofia seems to have considered her relationship with Vladimir over. She still wrote to him, but mentions of him in her diary virtually ceased.[18]

After the departure of Fufa and Sofia's apparent decision to put Vladimir and his problems out of her mind, Sofia was able to settle down to her mathematics again. But she was still working in isolation from other specialists. Only when Mittag-Leffler arrived in Paris with his bride at the end of May did this situation change.

Gösta Mittag-Leffler (1846–1927) was a handsome Swede of progressive tendencies whose love, respect and admiration for his teacher Weierstrass extended to cover Kovalevskaia, his teacher's favorite pupil. Mittag-Leffler was a capable mathematician, but his most important contributions to the history of his discipline were not his mathematical papers. He had a talent for mathematical analysis, but he had a genius for organization and for mathematical "intelligencing." He founded the journal *Acta Mathematica* and the Swedish Academy of Sciences mathematical institute which bears his name. In addition, he brought mainstream European mathematics to Scandinavia,

---

life, Sofia Vladimirovna (Fufa) applauded her mother's decision to send her back to Russia. (S. Vl. Kovalevskaia, "Vospominaniia," *Pamiati*, p. 145.)

[17]12 March 1882 letter from Clara Weierstrass in the IML.

[18]Diary entries for January 15–March 1, 1882, IML; Leffler, "Sonya Kovalevsky," pp. 195–96.

and Scandinavian mathematicians to the attention of the rest of Europe.[19]

Not the least of Mittag-Leffler's considerable achievements was his encouragement and sponsorship of Kovalevskaia, in the face of opposition by university administrators and some professors. He was a master tactician, and knew well the intricacies of academic politics. For the rest of her life he would advise Sofia on how best to conduct herself in her precarious position as first woman professor in modern Europe.

When Gösta arrived in Paris in May 1882, he immediately sought out Kovalevskaia. To his surprise he discovered that she had not made the acquaintance of French mathematicians, although she had been in the city for more than six months. With characteristic brusqueness, he waved aside her excuses and took her to visit the great French analyst Charles Hermite.[20]

Thanks to Mittag-Leffler's intervention, Kovalevskaia began to associate with her Parisian colleagues. Once she had been formally introduced to them, they were cordial enough, and invited her to take part in mathematical activities. On July 21, 1882, Kovalevskaia was chosen a member of the Paris Mathematical Society.[21] She got to know several of the best French specialists: Charles Hermite, Henri Poincaré, Emile Picard, Gaston Darboux and others, all of whom were helpful mathematically.[22] After her poor start, she worked efficiently in Paris, and was able to make significant progress on the light refraction problem.

Although Kovalevskaia had extensive professional contact with her French colleagues, at first she had little social contact

---

[19]André Weil, "Mittag-Leffler as I remember him," *Acta Mathematica*, 148 (1982), p. 12. Unfortunately, there is no full-length biography of Mittag-Leffler in English. For a short sketch, see Elfving, p. 72ff. Mittag-Leffler's manuscript memoirs of his Helsinki years are in the IML, as are thousands of letters from mathematicians all over the world.

[20]Mittag-Leffler, "Weierstrass et Sonja Kowalewsky," p. 183.

[21]*Bulletin de la société mathématique de France*, 10 (1881–1882), p. 254. She was nominated by the geometer C. Stephanos.

[22]Picard and Poincaré "in my opinion, are the most gifted of the new generation of mathematicians in all of Europe," Sofia wrote in her Autobiographical Sketch. (*Russian Childhood*, p. 221.)

with them at their homes. They visited her at her lodging, dined with her there, and invited her to dine with them in restaurants.[23] But before her husband's death, most of them did not invite her to meet their wives.

Kovalevskaia's ambiguous marital status and her choice of the typically male occupation of mathematician caused consternation among proper French matrons, who considered that she must be not quite respectable. Moreover, unattached Russian women generally had a bad reputation in stiff French society, because they were rumored to be political radicals and advocates of free love. This reluctance on the part of Sofia's Parisian mathematical associates to violate the wishes of their wives and invite her into their homes caused her ironic amusement and some bitterness.[24]

In spite of this lapse on the part of her professional friends, Kovalevskaia had an active social life during her time in Paris. The city had a large Russian and Polish revolutionary émigré community.[25] One of its leaders, P. L. Lavrov, at one time had been a colleague of General Korvin-Krukovskii at the artillery academy in St. Petersburg. He therefore had known Sofia's family for years. Through Lavrov and her sister Aniuta, Sofia soon made friends among the émigrés.

By far her best friend in Paris was the Polish revolutionary Maria Jankowska (later Mendelson), whom Sofia met in 1882 at Lavrov's apartment after Maria's release from Poznan prison. Jankowska was at first alienated by Kovalevskaia, because she appeared to be so avidly, superficially curious about Maria's experience in prison. During the course of the evening, however, Jankowska's opinion changed. Lavrov and Kovalevskaia began arguing about socialism. Sofia contended that the theory was too economically based, while Lavrov tried to convince her of the necessity of economic considerations. Both of them

---

[23]Diary entries from June 1882, IML.

[24]Kovalevskaia, 1883 letters to Mittag-Leffler, IML; Kovalevskaia, undated [1883] letter to Georg Vollmar, *VIP* 1951, pp. 265–66.

[25]For an idea of the life of the revolutionary émigrés, with its poverty, camaraderie, and factional bickering, see L. G. Deich, *Russkaia revoliutsionnaia emigratsiia* (Petersburg: Gosizdat, 1920); Lavrov's letters to E. S. Shtakenshneider in *Golos minuvshego*, No. 7–8 and 9 (1916); Pomper, *Peter Lavrov and Sergei Nechaev* (New Brunswick: Rutgers University Press, 1979).

spoke idealistically of everyone's rights to equal happiness, and about the revolution (for Sofia a bloodless one) which would bring about this wonderful state of affairs.

To Jankowska, who had just spent time in a tsarist prison, "it all sounded very beautiful and very indefinite." Still, she was impressed by Sofia's sincerity and obvious desire to understand socialism, and was drawn to her as a person: "everything about her attracted and interested me." The two women, both in their early thirties, both intelligent and lively, both with some political consciousness, became good friends. Sofia loaned her passport to Maria for illegal trips to Russian Poland, and helped her friend's political activities in other ways. Maria gave up her Paris apartment to Aniuta during the latter's final illness in 1886–87, and provided a place for Sofia to stay on her trips to France.[26]

Jankowska-Mendelson was not Kovalevskaia's only friend during that year in Paris. As always, when Sofia set out to be charming and actively sought social interactions, she soon gathered a crowd of friends and admirers. One trait which made her popular was her ability to focus her attention exclusively on the person, male or female, with whom she was talking. Maria Jankowska-Mendelson recalled that:

> Each of Sofia's numerous friends preserved in his or her memory a different image, because to each one she presented herself in a completely different light. But there was not the least falseness in this; it was just that her rich nature gave to the person who interested her at that moment exactly what seemed to her to suit that person. Her interest in the given personality was exclusive; it did not permit her to involve herself with others at the same time.[27]

This absorption in one person at a time is reflected in Sofia's diary entries for 1882.[28]

---

[26]M. V. [Jankowska] Mendelson, "Vospominaniia o S. V. Kovalevskoi," *Sovremennyi mir*, No. 2 (1912), pp. 137–39 and *passim*.

[27]Mendelson, p. 136.

[28]Kovalevskaia, 1882 diary, IML.

There were times when Sofia persuaded herself that her exclusive absorption in some individual meant that she was falling in love. This happened twice in 1882—with the German Social Democrat Georg Vollmar, and the radical Polish mathematician-poet Joseph Perrot (also spelled Perott and Perrott).

Kovalevskaia met Vollmar at Aniuta's on March 12, and was charmed by him. She discussed politics with him, went to socialist meetings in his company, and carried on arguments with herself about whether Vladimir deserved her faithfulness.

When Vollmar returned to Germany, Sofia seriously considered joining him there. I am so lonely, she mused in her diary. What would be the harm of an association with Vollmar? Only one consideration prevented her from abandoning respectability: Sofia was worried about Fufa. She seems to have reluctantly come to the conclusion that Fufa was more important than Vollmar. She thought no more about a liaison with the German, although they remained friends until her death.[29]

Sofia's dealings with Joseph Perrot were at once more complicated and more superficial. They studied together, he told her about his mistress and his affairs, and she allowed his sister Zinaida to live with her in Paris. Moreover, there are some indications in her diary and correspondence that she was helping him with an unspecified political project. Possibly, she was handling the finances for his radical organization.

Kovalevskaia thought about Perrot often, and went so far as to write passionate love letters to him. She did not mail them, and indeed, in one of them she admits that she is, so to speak, making him the basic material out of which she is fashioning her ideal fantasy lover. By contrast, Perrot's emotions toward her were those of mild friendship at best. His letters to Sofia are often rude and arrogant. They convey the impression that he used her politically and mathematically (virtually demanding her constant aid in his research), if not emotionally.[30]

---

[29]25 March through 28 May 1882 diary entries in the IML. See also references to Vollmar in Mendelson, *passim*, and Mittag-Leffler's letters to Weierstrass, IML.

[30]See veiled political allusions in Joseph and Zinaida (Zina) Perrot's 1882–83 letters to Kovalevskaia, her October 1882 telegram and letter to Mittag-Leffler, and Perrot's 11 November 1882 note to Mittag-Leffler, all in the IML.

Interestingly enough, in later life Perrot was not above boasting about his association with the famous Sofia Kovalevskaia. One American mathematician, secretary of the Mathematical Association of America R. C. Archibald, went so far as to ask Mittag-Leffler about Perrot's claims. "I have met the man and can scarcely credit that even when he was much younger he could be at all attractive to such a brilliant woman," Archibald wrote. Mittag-Leffler indignantly refuted the rumor, writing that while Sofia knew Perrot, her affections lay elsewhere.[31]

There is no doubt that Perrot knew Kovalevskaia fairly well, and had reason to be grateful to her. She could be extraordinarily patient and kind to those in misfortune, or to those who she believed were trying to better themselves intellectually. When Perrot told her that his sixteen-year-old sister Zina wanted to study abroad, but had been forbidden to leave Russian Poland by her parents, Sofia calmly sent her passport for the girl's use. When Maria Jankowska-Mendelson remonstrated with her on the danger of handing over her papers to anyone who asked her, Sofia replied with what her friend considered a characteristic blend of romanticism and decisiveness:

> . . . is it really possible not to stretch out one's hand, to refuse to help someone who is seeking knowledge and cannot help herself reach its source? After all, on woman's road, when a woman wants to take a path other than the well-trodden one leading to [traditional] marriage, so many difficulties pile up. I myself encountered many of these. Therefore I consider it my duty to destroy whatever obstacles I can in the paths of others. According to her brother, this girl has unusual capabilities in the exact sciences. Who knows, maybe she'll become a prominent scholar![32]

Even after Zina arrived, and proved to be more interested in the Parisian social scene than in scientific study, Sofia did not get angry or abandon her. She allowed Zina to stay in her

---

Sofia's fantasy letters, addressed to "Don José" (Perrot's Spanish mistress' name for him) are also in the IML.

[31]R. C. Archibald, 23 July and 22 December 1922 letters to Mittag-Leffler, IML. Mittag-Leffler specifically mentions Maksim M. Kovalevskii, who will be discussed below.

[32]Mendelson, p. 140.

rooms, and attempted to tutor her in mathematics. Her fantasy of helping a burgeoning female Newton or Gauss was not to be. But Sofia acknowledged her disappointment with Zina with ironic self-mockery rather than anger. Jankowska-Mendelson observed that Kovalevskaia "was extremely sweet every time she returned to reality from one of her golden dreams: it was as if she laughed at herself, recognizing that she knew neither people nor the world."[33]

This recognition of her own innocence and naïvete was undoubtedly part of the reason for Sofia's interest in the friendship of more sophisticated and committed political activists. Her letters to Jankowska-Mendelson and Georg Vollmar, whom she visited every time she went to Berlin, indicate that she felt guilty about her lack of knowledge of and commitment to radical politics. She envied her friends their involvement in active political work, gave them her passport, accepted coded letters, and helped out in other small ways that entailed some risk. Yet she could not bring herself to come right out and join them. Mathematics was more important to her.

In a May 4, 1882 letter to Georg Vollmar, Kovalevskaia wrote that after becoming so close to revolutionary circles in Paris, she sometimes regretted choosing such an abstract discipline as mathematics. In her youth, she said, she had felt that the only way for an honest, right-thinking person to live honorably in a country with the unjust social order of Russia was to devote oneself to science, trusting science and education to make everything right eventually. To be sure, she had always considered herself a socialist, at least in principle. "But I ought to confess to you that the resolution of the social question seemed to me so far away and murky, that completely devoting oneself to the issue did not seem worthwhile for a serious scholar capable of doing something better."[34]

Now, when Kovalevskaia was close to people like Lavrov, Jankowska-Mendelson, and Vollmar, who made sacrifices and in some cases devoted their lives to changing the political order, she often had the disturbing feeling that what she did was useful only for a small number of specialists. This was particu-

---

[33]*Ibid.*, pp. 140–41.

[34]Letter in *VIP 1951*, p. 263.

larly shameful, she thought, because now was the time "when everyone has the responsibility to devote their best energies to the cause of the majority."[35] These political guilt feelings were to trouble her for the rest of her life.

While Sofia worked on her light refraction paper and socialized with her Russian and Polish émigré friends, Vladimir was moving headlong toward disaster. In his effort "to combine service to geology with service to Mammon," as Sofia had phrased it, he managed to serve neither. His powers of concentration were so diminished that it took him several hours to prepare one page of notes for his class. He delivered one mediocre lecture after another, and even started missing classes to do work for the Ragozins.[36]

His association with the company was not going well either. The warnings Sofia sent him finally took hold, and Kovalevskii began to have doubts about the honesty of the company's main shareholders, Viktor and Leonid Ragozin. He made several inept attempts to take over the board of overseers by forcing stockholder votes, but only succeeded in having himself removed from his position as director. He could not leave the company, though, because he was too much in debt to the brothers and because Viktor Ragozin apparently threatened him with legal action if he departed.[37]

Kovalevskii tried to persuade his brother to buy more stocks, perhaps to appease the Ragozins, or perhaps reasoning that if the company was dishonest, Aleksander might as well take advantage of it. After an initial refusal, the older Kovalevskii agreed to mortgage his house for funds. Fortunately, his old friend, the physiologist I. I. Mechnikov, who was charged with

[35]*Ibid.*, p. 264.

[36]November 1882 letters from Vladimir to Aleksander in Davitashvili, p. 352ff.; Shtraikh, *Sestry*, p. 281ff.; Reznik, *Vladimir Kovalevskii*, p. 307ff. According to the zoologist V. N. Lvov, Kovalevskii was an awful lecturer. The only time he excelled was once when, coming in late, he brought the wing of a dead crow he'd found on the street. Forgetting his prepared lecture, to the relief of all his students, he delivered a brilliant impromptu discourse on the evolution of the ability to fly. That lecture seems to have been the only glimpse his Moscow pupils had of the spontaneous genius Kovalevskii used to display in his student years abroad. (Borisiak, p. 57, n. 4.)

[37]See letters in *N N*, p. 375ff., and notes on p. 420.

carrying out the transaction, opposed Aleksander's plans. Mechnikov feared that the financial deal would lead to bankruptcy, and resisted turning over the money to Vladimir. Thanks to this intervention, Aleksander took some losses but was not financially ruined when the Ragozin crash came.[38]

From the time Sofia went abroad in 1881 until his suicide in April 1883, Vladimir's sentiments about his relations with his wife vacillated so wildly that she had trouble keeping track of where she stood in his affections at any point. At times he wrote tenderly, asking her advice and saying he could hardly wait to see her. Then, before she had a chance to reply, she would receive a letter addressed to the "much respected Sofia Vasilevna," a formal title specifically designed to enrage her.[39]

In his letters to Aleksander, Vladimir sometimes blamed Sofia for the state of affairs in which he found himself, and regretted tying his future to hers in 1868. In May 1882 he formed the notion of returning to Sofia the 50,000 rubles that were her inheritance from her father: the 20,000 rubles Vladimir had borrowed with her consent, and the 30,000 rubles he had appropriated at various times for further speculations. This was to be a grand gesture, accomplished through a "council" consisting of Aleksander Kovalevskii, Iulia Lermontova, Maria Bokova-Sechenova, and Ivan Sechenov. Effectively, it was to be Kovalevskii's dramatic renunciation of his life with Sofia, about which he now said: "And under this regime I wasted fourteen years, for God's sake!"[40]

The intent was clearly to hurt Sofia, since she had previously insisted that half of her inheritance was Vladimir's. He succeeded in his intention to wound her, but could not return the money, which he did not have. In fact, the whole gesture was just one more symptom of Kovalevskii's mental deterioration. But it was particularly unpleasant to Sofia, because she thought that they had parted as friends after Vladimir's last visit. She complained to her brother-in-law:

> What all this means, I positively do not understand. Even without this the present situation is difficult, and to have on top

[38]Reznik, *Vladimir Kovalevskii*, pp. 299–300.

[39]Letters in Borisiak, p. 59; Shtraikh, *Sestry*, p. 284.

[40]7 May 1882 letter to Aleksander, in Davitashvili, p. 349.

of it such a surprising attitude on the part of Vladimir. . . . It is even stranger for me, because before he left I could not have presupposed anything like this. After all, until the last minute I intended to travel with him. I remained only because of economy; we parted as friends. . . . What exactly does Vladimir hope for?—I positively cannot understand [him].[41]

Kovalevskaia's exasperated query would be repeated many times in the next few months, and would get no answer. Vladimir was far past the point when he could have explained his actions in anything approaching rational terms. In fact, he was already contemplating suicide.

Kovalevskii's letters to his brother and Iulia Lermontova in the last months of his life are pathetic. Increasingly, he blamed himself for his inadequacies, and felt that he understood Sofia's and Aleksander's disappointment in him. "Only now have I come to realize how much good, pure love was wasted on me by such people as you and Sofa," he wrote in December 1882. Over the winter holidays he visited his brother in Odessa, and wrote to Iulia about his state of mind: "I thought for a while that maybe some kind of change for the better would take place [in my mood] but everything remains just as gloomy as it was before. . . . It's simply painful and frightening to look at Fufa and think about the future . . ."[42]

Kovalevskii began to arrange his affairs, as far as they could be arranged given the tragic confusion he had made of them. He wrote his brother a farewell letter on February 1, 1883, but did not mail it; it was found among his things after his death. For two months he postponed his decision. He lived from day to day, hoping that his concentration would return sufficiently for him to resume scientific work, and that the danger of a trial of the Ragozin company directorate would pass. Neither of these things happened. Moreover, Kovalevskii learned that the

[41][Late spring 1882] letter in Shtraikh, *Sestry*, pp. 282–83.

[42]7 December 1882 letter to Aleksander and 16 January 1883 letter to Iulia in Davitashvili, pp. 354–57. This letter to Iulia and others to her around the same time indicate the possibility that Iulia actually knew of Vladimir's intention to commit suicide. Perhaps she did not really think he would go through with his plan.

work he had submitted for his Russian doctorate was not considered good enough for a thesis.

The paper, on fossil formations in France, dated from Kovalevskii's time abroad before 1874, but it had not yet been published and so could still have been used for a dissertation. Although it was not the most impressive of Vladimir's works, nevertheless it certainly compared favorably to other doctoral dissertations defended at that time in Moscow. That it was not accepted seems to have been sheer prejudice on the part of Moscow geologists. Vladimir's foreign training, his arrogant championship of all German science, and his peculiar personality angered the Moscow specialists too much for them to be objective about his work.

The blows were too many, and Vladimir's mental balance was too fragile to sustain them. On the night of April 27/28, 1883, he committed suicide by inhaling an entire bottle of chloroform.[43] His last message to Sofia and Aniuta, enclosed in his farewell letter to his brother, was short but tender:

> Write to Sofa that my constant thoughts were of her and of how much I am at fault before her, and how I spoiled her life, which without me would have been bright and happy. My last request to Aniuta is that she look after her and little Fufa; she [Aniuta] is the only one now capable of doing that, and I beseech her to do so.[44]

Kovalevskii wrote a one-sentence goodbye to Iulia Lermontova: "Forgive me, I cannot do otherwise." He also wrote a farewell to Maria Bokova-Sechenova, in which he asked her to consider taking Fufa to raise. Finally, he requested his acquaintance, the jurist A. I. Iazykov, to try to clear his name in connection with the Ragozin affair. In his letter to Iazykov, Vladimir stated formally that although he killed himself because of the upcoming Ragozin trial, he swore that he was

---

[43]"Strannaia smert'. Samoubiistvo V. O. Kovalevskogo," *Novoe vremia*, 21 April 1883.

[44]*N N*, p. 385.

innocent of any wrongdoing.[45] His only goodbye to Sofia was the few lines cited above.

Kovalevskii was sincerely mourned by specialists throughout Western Europe and America. His productive scientific career was short. In fact, it spanned only the years of his time abroad with Sofia. But he is considered one of the founders of the science of evolutionary paleontology, and his work on fossil horses was the point of departure for all future study of the subject.[46]

When Kovalevskaia heard the news of Vladimir's suicide, she could not bring herself to eat and refused to see anyone, including Maria Jankowska-Mendelson and the doctor Maria sent. She blamed herself for her husband's death, and was convinced that she could have prevented it if she had been in Moscow.

It is possible that Sofia's reaction to Vladimir's death was so extreme because she felt guilty about not caring more. She had come to look upon her marriage as a burden, and considered herself bound to Vladimir more from a sense of duty and responsibility than from love. Perhaps she tried to starve herself as a punishment for feeling relieved by her husband's death.

Five days after she began her fast, Sofia fainted from hunger. The doctor was able to force-feed her while she was unconscious, and her protegée Zina Perrot, who was still living with her, tended her during her recovery.

The next day Sofia regained consciousness, and began working intensively on mathematics. In fact, it was at this time that she finished her work on the light refraction problem. It seemed as if she could not trust herself to think of anything other than work, and had deliberately blanked out all memory

[45]All letters are quoted in Sofia's October 1883 letter to Aleksander Kovalevskii in *VIP 1951*, p. 269. Sofia found the letters when she was going through Vladimir's effects to find evidence of his innocence in the Ragozin conspiracy trial.

[46]For a summary of evaluations of Kovalevskii's work by Western scientists, see Borisiak, pp. 118–33 and *passim*; Davitashvili, p. 376ff. and *passim*; Vucinich, p. 118ff. and *passim*. There is a 1903 letter in the IML from the Curator of the Belgian Royal Museum of Natural History, Louis Dollo. Although it was written 20 years after Vladimir's death, it is clear that Dollo still mourned Kovalevskii and valued his work.

of what had caused her collapse. When Jankowska-Mendelson saw her that day, she thought that "her eyes were reminiscent of those of an obedient, intelligent child."[47]

Kovalevskaia made a slow recovery from her weakness after Vladimir's death. But finally, in early summer, she felt well enough to travel to Russia to put her husband's things in order. Because he had had no relatives in Moscow, and was bankrupt, Kovalevskii had been given a pauper's funeral by the city. His private papers, fossil collections and books only escaped being sold to pay for the burial through the intercession of the Moscow University rector, N. S. Tikhonravov. The rector locked up Vladimir's office and laboratory and refused to let the authorities inside, insisting that everything within was university property.[48]

Thanks to Tikhonravov's intervention, Kovalevskaia was able to gather all Vladimir's papers pertaining to the Ragozin affair, and take them to the prosecutor's office. The prosecutor at first seemed disposed to accuse Vladimir posthumously along with the Ragozins, who did their best to implicate Kovalevskii as much as possible. Happily, after reading through the papers Sofia gave him, the prosecutor came to the charitable conclusion that "V. O. had been carried away [by the affair], but was an honest man."[49]

That summer and fall Kovalevskaia remained in Moscow to straighten out her husband's finances, clear his name, and arrange for an inscription for his grave.[50] At the same time, she was negotiating with Mittag-Leffler and Weierstrass about a position in Stockholm, where Mittag-Leffler was now teaching. On her way to Russia, she had stopped in Berlin to show her completed work on the light refraction problem to her former teacher. Weierstrass was satisfied, and encouraged her to write up her results for presentation. At the end of August Kovalev-

---

[47]Mendelson, pp. 147–48.

[48]Letter from Lermontova, *N N*, pp. 392–93.

[49]Kovalevskaia, letter to Aleksander Kovalevskii, *VIP 1951*, pp. 269–70.

[50]The monument inscription Kovalevskaia chose was: "And suddenly your soul found peace/the like of which joy/was never found in earthly life/the peace of non-existence." (Davitashvili, p. 371.)

skaia delivered her paper at the Seventh Congress of Natural Scientists and Doctors in Odessa, where it was well received.[51]

Sofia was pleased at her reception by Russian scientists, but her finances were running low, and she needed to arrange a position so that she could support herself and Fufa. She wrote to Jankowska that she had resigned herself to a post in the lowest grades of a girls' *gymnasium*, and intended to go to St. Petersburg to find such a place.[52]

Sofia's sister Aniuta spent part of the year in the Russian capital, where she lived with their old comrade Zhanna Evreinova. Aniuta did translations and wrote stories, while Zhanna edited the progressive "thick journal" *Severnyi vestnik* (Northern Messenger). After Vladimir's death, the thought of living with these old friends must have been appealing to Sofia.[53]

Meanwhile, Vladimir's death had caused unmitigated relief in mathematical circles. As a widow Sofia would encounter fewer social obstacles to her mathematical career than as a single or married woman. In Europe in the nineteenth century, widowhood was eminently respectable, and at the same time allowed for much greater independence in a woman's life than the single or married state.

Weierstrass wrote to Mittag-Leffler that with Vladimir's death, his objection to Sofia obtaining a post had been removed. He urged Gösta to proceed with arrangements in Stockholm. He said he was well-pleased with Sofia's refraction work, and gave the Swede a full description of the results.

Moreover, Weierstrass gave Mittag-Leffler a detailed explanation of the circumstances of Kovalevskaia's marriage. This was not an idle indulgence in gossip on the part of the great German mathematician. Rather, he knew that the appointment of a woman would not be achieved without acrimonious de-

---

[51]Kovalevskaia's paper actually had an error in it, which neither she nor Weierstrass noticed. The Italian mathematician Vito Volterra discovered it a few months after her death in 1891. See Vito Volterra, 3 June 1891 letter to Mittag-Leffler, IML.

[52]Mendelson, p. 149.

[53]Undated letter to Aniuta, *VIP 1951*, pp. 266–67.

bate. He wanted Gösta to be fully prepared to deal with snide remarks and innuendoes.[54]

Mittag-Leffler was an excellent politician, and for years he had felt that Kovalevskaia would be the ideal colleague in a small mathematics department. He and several other professors made a concerted assault on the administration of the four-year-old Stockholm University. By skillful maneuvering, they obtained a temporary, provisional post for Sofia before reactionary faculty and conservative elements in Swedish society could martial their forces.

Considering the fact that Kovalevskaia was the first woman in modern times to be appointed to a position in a European university, approval of her went relatively smoothly. Gösta Mittag-Leffler was not deceived by this first victory, however. As he remarked, "the true difficulties will come later."[55]

Kovalevskaia received word of the Stockholm University post from Weierstrass in early September 1883.[56] The offer—as *privat docent*—was a common one given to people right out of graduate school. Sofia was to be on probation for a year, during which time she would receive no salary and no official status. Her pupils would pay her by private arrangement, and her situation would come up for review at the end of the academic year.

Given the fact that Kovalevskaia was to be the only woman in Europe to have a university-level teaching post, it is not surprising that Sofia felt awed by the offer. She wrote to Mittag-Leffler:

> I cannot tell you how grateful I am to you for the friendship you have always shown me, and how happy I am to be able to enter a career which has always been the cherished object of my desires. At the same time I feel I ought to tell you that in many respects I feel but little fitted for the duties of 'docent', and at times I so much doubt my own capacity that I feel that you, who

[54]Weierstrass, 8 and 27 August 1883 letters to Mittag-Leffler, IML.

[55]Mittag-Leffler, 20 October 1881 letter to Weierstrass, IML; Mittag-Leffler, "Weierstrass et Sonja Kowalewsky," p. 192.

[56]Weierstrass, 27 August 1883 letter to Kovalevskaia, *Briefe/Pis'ma*, pp. 103–06. If Mittag-Leffler also wrote to Kovalevskaia about her post, the letter does not appear to have survived.

have always judged me leniently, will be quite disillusioned when you find, on nearer inspection, how little I am really good for.[57]

Kovalevskaia wrote in a similar vein to Weierstrass. He sent back a long, comforting, supportive letter. He knew she was competent, he said. There was no need to be nervous. He advised her to subdue her fears and assume the post as soon as possible.[58]

This self-abnegation on Kovalevskaia's part was in one sense surprising, since a university position represented the summit of her dreams. Yet in another sense it was perfectly understandable. It is likely that Sofia's reluctance was the same as that of many women who, in the modern terminology, have internalized the prejudice against them to the point where they sometimes really believe that they are inferior.

Elizaveta Litvinova described those feelings with the empathy of one who had experienced them herself:

> This attitude is the exclusive property of women who blaze completely new paths. The prejudgment against women's capabilities for intellectual work, as all prejudices against which we have to fight, lives not only in those around us, but also in ourselves. The exact same confusion would be felt by any talented woman, but the thought that he was not well enough prepared to assume the duties of docent would not even pass through the mind of the most mediocre man.[59]

Fortunately, Kovalevskaia was able to overcome for the moment her misgivings about her supposed unworthiness. With the help of Mittag-Leffler and Weierstrass, she convinced her-

[57]Kovalevskaia, 28 August 1883 letter to Mittag-Leffler, IML, published in Leffler, "Sonya Kovalevsky," p. 205.

[58]Weierstrass, 12 September 1883 letter to Kovalevskaia, *Briefe/Pis'ma*, pp. 106–08.

[59]Litvinova, p. 58. Litvinova went on to remark that this internalization of inferiority led Kovalevskaia to look upon herself exclusively as Weierstrass' student. Her excessive humility caused later commentators to exaggerate her dependence on her former teacher.

self that she was indeed capable of teaching at the college level, and set about the arduous task of winding up her affairs in Russia. Fufa was left with Iulia Lermontova until such time as her mother was permanently settled in Sweden. The physicist Lamanskii was delegated Sofia's business representative in St. Petersburg, Sofia's share of the Palibino furniture was put into storage, and Vladimir's affairs were entrusted to relatives.[60]

On November 15, 1883, at the age of thirty-three, Kovalevskaia left St. Petersburg for Stockholm, by boat via Finland. Behind her was the struggle to return to mathematics after five years away from research, the tragedy of her husband's suicide, and the pain of having to fight to clear his name after his death. Behind her also, for the time being, were her doubts about her capabilities.

Kovalevskaia was aware that her appointment to the position in Stockholm was an event of importance for women throughout Europe. If she could succeed in regularizing her post, her case would serve as a valuable precedent.[61] She was determined not to fail, to open the road of university teaching finally and completely to women in Europe.

The first years would be difficult. Kovalevskaia would have to contend with the prejudice of her university colleagues, the antagonism of reactionaries, and material deprivation. She would be separated from her daughter, her Parisian friends, the familiar society of St. Petersburg. She would have to learn a new language, lecture on mathematics for the first time in her life, and prove herself to the university authorities.

Kovalevskaia would be lonely. She would often feel stifled by the cool respectability of Swedish life. Yet she would do

---

[60]1883 and 1884 business correspondence with S. A. Lamanskii and others in the IML. Sofia's younger brother Fedia sold Palibino around this time to cover his gambling debts, so Sofia was obliged to help him financially. (Sofia Vladimirovna Kovalevskaia, "Vospominaniia," *VIP 1951*, p. 368.)

[61]The cases of several late Renaissance Italian women who had had university chairs were not widely known at this time, and therefore were not considered as precedents. S. Nikitenko has interesting archival information on these women in his "Zhenshchiny-professora Bolonskogo universiteta," *Russkaia mysl'*, No. 10 (1883), p. 256ff.

some of her best creative work, literary as well as mathematical, during her Stockholm period. And the years would be crowned by triumphs in her scientific career the likes of which she had never expected: an extraordinary (i.e., assistant) professorship at Stockholm University, and the Prix Bordin of the French Academy of Sciences.

ten

"A PRINCESS OF SCIENCE"

Kovalevskaia arrived in Stockholm on November 17, 1883 to take up the position she had worked so hard to achieve—university teacher. She stayed with the parents of the mathematician Mittag-Leffler until she could find an apartment of her own. The Lefflers—Gösta, his wife Signe, his sister Anna Carlotta, his brother Fritz, his parents, uncles, aunts, and cousins—became Sofia's substitute family.

In general, the people she met in those first few weeks were friendly and eager to make her feel at home. The young women were "even ecstatic," as Sofia wrote to her brother-in-law, and welcomed her as a fellow fighter for women's emancipation. One of the progressive Stockholm newspapers heralded her coming in extravagant terms:

> Today we must inform you *not* of the arrival of some commonplace prince or similarly highly-placed but totally ignorant individual. No, rather it is a princess of science, Mme. Kovalevskaia, who has honored our city with her visit and will be the first woman *privat docent* in all of Sweden.

Sofia was pleased by the flattery. But she added the ironic comment: "Look at that! I've been made into a princess! It would have been better if they had given me a salary."[1]

Although Mittag-Leffler, most of his colleagues at the university, and progressive Swedish society in general were friendly

---

[1]December 1883 letter to Aleksander, *VIP 1951*, pp. 275–76.

to Kovalevskaia, it was not long before she discovered that opposition to her had not been silenced by her appointment. As Elizaveta Litvinova put it, Sofia was an "apple of discord" between the "Young Sweden" faction, whose representatives included Mittag-Leffler and several other professors at Stockholm University, and the academic reactionaries based chiefly at the established universities of Uppsala and Lund.

Stockholm University was a new institution. It had been set up by progressives in the capital specifically as an alternative to the conservative older universities of Uppsala and Lund. It was financed largely by private contributions, and had admitted women students on an equal basis with men from the day of its founding.

The Stockholm liberals strongly believed in the equality of women, and in the necessity of "liberating science from the conservatism of the old universities."[2] Most Uppsala and Lund professors, on the other hand, insisted on traditional beliefs and methods of instruction, and were furious that even such a humble position as Sofia's had been given to a woman.

Kovalevskaia was fascinated by this "lively battle between the new and the old," as she called it in a letter to S. A. Iurev, editor of the journal *Russkaia mysl'* (Russian Thought).[3] She loved controversy, and this type of progressive *vs.* reactionary debate was much to her taste. She was even amused rather than upset when she found out that notices of her lectures, put up by students in their union in Uppsala, had been defaced by the professors.[4]

Kovalevskaia was determined to learn Swedish as soon as possible, so that she could participate in the intellectual and political life of Stockholm. For the first few weeks, she did virtually nothing but study the language. Within a fortnight, much to the surprise of Mittag-Leffler and his wife, Sofia was able to converse sufficiently well so that she did not embarrass herself at receptions given in her honor.[5]

[2]Litvinova, pp. 60–1.

[3]December 1883 letter in *VIP 1951*, p. 272.

[4]December 1883 letter to Maria Jankowska-Mendelson, *VIP 1951*, p. 274.

[5]Leffler, "Sonya Kovalevsky," p. 210. Anna Carlotta Leffler noted that Kovalevskaia spoke all foreign languages in the same way: quite fluently but

Once Kovalevskaia had picked up enough Swedish, she followed the conservative/progressive debates in the popular press with interest. She was particularly drawn to the new generation of Scandinavians who wrote on social themes—August Strindberg, Henrik Ibsen, and Anna Carlotta Leffler. (Leffler, while virtually unknown outside of Scandinavia today, was often compared to Ibsen and Strindberg during her lifetime.) With some amusement Sofia learned that Mittag-Leffler's sister Anna Carlotta was considered "a great revolutionary" in Sweden.[6] To Sofia, who had recently come from the Polish and Russian revolutionary émigré circles in Paris, and was steeped in the radicalism of her sister Aniuta and her friend Maria Jankowska-Mendelson, Anna Carlotta's gentle social rebellion seemed lukewarm in comparison. Sofia set about influencing the Swedish woman's writing to conform to her more sweeping political standards.

As usual, when Kovalevskaia decided to accomplish something, she lost no time. Before their first meeting was over Leffler found herself telling Sofia the plot of the play she was working on, "in far greater detail and breadth than I had ever been conscious of intending." For Leffler:

> This was the commencement of the great influence she exercised on my writings afterward. Her power of understanding and sympathizing with the thoughts of others was so exceptional, her praise when she was pleased so warm and enthusiastic, her criticism so just, that, for a receptive nature like mine, it was impossible to work without her approbation.[7]

Sofia was proud of her strong influence on Anna Carlotta's writings. Kovalevskaia once said that Leffler never would have written *Ideal Woman* and her other pre-1884 works if Sofia had been around to stop her. Their collaboration was intense, and culminated in the two joint plays *The Struggle For Happiness: How It Was* and *How It Might Have Been* (1887). Then, Anna

---

with errors while she lived in a country, very poorly when she had not used the tongue for a while.

[6]December 1883 letter to S. A. Iurev, *VIP 1951*, p. 273.

[7]Leffler, "Sonya Kovalevsky," pp. 208–09.

Carlotta began to feel that Sofia was stifling her, and went abroad so that she would be able to write in her own way, rather than in the way Sofia dictated.[8]

When Kovalevskaia arrived in mid-November, Christmas holidays were about to begin. She thus had ample time to settle into life in Sweden before she started teaching. She searched for an apartment, made new friends, and learned about her enemies. She found much to occupy and amuse her in the two months before her lectures began. But she was lonely, and still felt sorrow at Vladimir's death, as entries in her diary attest.[9]

Mittag-Leffler had advertised Kovalevskaia's lectures beforehand. There were to be twelve students—"that is, everyone who is studying higher mathematics," as Sofia wrote to her brother-in-law. She was nervous about teaching for the first time, and had prepared several classes in advance. The subject, partial differential equations, was one which she knew well, and in which she had done important research.[10] As Sofia wrote to Aleksander Kovalevskii: "I am very glad about that. It seems to me that it will be impressive if a woman, beginning her first course of lectures, can speak in part about her own findings in that subject [on which she lectures]."[11]

On January 30, 1884, Kovalevskaia gave her first lecture, in German, on partial differential equations. The auditorium was full; people were aware of the historic nature of the occasion. Not only the twelve enrolled students, but also other students, professors, university officials, and interested citizens came to see the "princess of science" begin her teaching career. Sofia was nervous, and stumbled at first, but finished her talk to

---

[8]*Ibid.*, pp. 209, 256–57. Possibly, it was Anna Carlotta's resentment at Sofia's tampering with her works that caused Leffler later to take certain liberties with the details of Kovalevskaia's life.

[9]See January 1884 entries in *VIP 1951*, pp. 177–78.

[10]One of Kovalevskaia's doctoral dissertations contained a result basic to the theory of partial differential equations, known today as the Cauchy-Kovalevskaia Theorem. See Chapter XIII for further details.

[11]October 1883 letter to Aleksander, *VIP 1951*, p. 271.

applause. It was clear even from the first class that she would be a good lecturer.[12]

Although her teaching was successful, Sofia was finding life in Stockholm more lonely and humdrum than she had thought it would be. She found no Swedish equivalent of Maria Jankowska-Mendelson or Iulia Lermontova or her sister Aniuta; no friend to whom she could express her thoughts. Her relationship with Anna Carlotta did not satisfy the need for a confidante, although it came the closest. Leffler was too aloof, too afraid of submerging her own weaker personality in that of her friend, to be a completely congenial companion for Sofia.[13]

As a group, the people of Stockholm with whom Sofia came into contact at first, even those of the younger generation, were too conventional for her taste. She did not like the way they gossiped, and were so inquisitive about her personal life. She complained to Weierstrass and Maria Jankowska-Mendelson about that tendency, and begged Maria to visit her in Stockholm to relieve her feelings of spiritual isolation.[14]

Indeed, then as now, many Russian and other East European émigré intellectuals tended to feel lonely and cut off when abroad. They looked upon their Western counterparts as superficial and/or cold, and considered the Westerners' concerns trivial and bourgeois. Kovalevskaia was no exception.

The language barrier was also a problem for Sofia. She quickly learned enough Swedish to make herself understood on the street, was soon able to read without difficulty, and after the first semester conducted her lectures in Swedish. In superficial social situations she did not need a large vocabulary and could always make her meaning clear. However, she longed to be able to speak Russian, for it was only in that language that Sofia felt completely herself. She sometimes complained to Anna Carlotta:

[12]Mittag-Leffler, 18 February 1884 to Weierstrass, IML.

[13]See diary entries 30 January through 21 February 1884, *VIP 1951*, pp. 178–80; Leffler, "Sonya Kovalevsky," p. 211ff., and *passim*.

[14]*Briefe/Pis'ma*, pp. 109–10; Mendelson, p. 154ff.

I can never quite express the delicate *nuances* of thought. I have always to content myself with the next best expression, or say what I want to say in a roundabout way. I never find the exact expressions. That is why, when I return to Russia, I feel released from the prison in which my best thoughts were in bondage. You cannot think what suffering it is to have to speak always a foreign language to your friends. You might as well wear a mask on your face.[15]

In the liberal-bourgeois atmosphere of Stockholm university circles, where her radical political convictions as well as the language difficulties separated her from her companions, Sofia grew more interested in underground movements rather than less. Her correspondence with Maria Jankowska-Mendelson, Petr Lavrov, and Georg Vollmar was laregely on political topics, and she followed reports of revolutionary activity in Poland, Russia, and Germany. She began inquiring about the socialist movement in Sweden, and sounded out acquaintances on their theoretical opinions.

Soon after her arrival in Stockholm, Kovalevskaia wrote a detailed letter to Lavrov, asking him to consider the advantages of making the city a sort of underground railway stop for Russians and Poles fleeing the tsarist police. Sweden shared a common border with the Russian empire (which then included Finland), she pointed out. Moreover, there were several leading citizens who seemed to have some theoretical sympathy with socialist ideas. She had had conversations with prominent Swedes who openly voiced such sentiments, but "I do not yet claim to say positively how much this interest and sympathy is sincere, and how much role is played by the desire to say something congenial to the convictions my acquaintances assume in me."[16]

However much Kovalevskaia might have felt bored and isolated by life in Stockholm, and might have wished it could be enlivened by more revolutionary politics and a constant

[15]Leffler, "Sonya Kovalevsky," p. 211.

[16]December 1883 letter, *VIP 1951*, p. 277. There are two letters from Lavrov to Kovalevskaia in the IML. Lavrov does not seem to have been enthusiastic about Sofia's idea of using Stockholm as an underground railway stop.

stream of Russian exiles, she realized that the university there represented her only opportunity to teach in an institution of higher education. She was committed to teaching and to mathematical research, and felt grateful to Mittag-Leffler and his colleagues for giving her the chance to prove herself in this field. At the end of her first semester she wrote to Mittag-Leffler: "I shall never be able to express or show all the gratitude and friendship I feel for you. It is as if I had found in Sweden a new fosterland and family at the moment when I most needed them."[17]

Moreover, Kovalevskaia was finding the Swedish environment conducive to mathematical work. As mathematicians know, one of the important criteria for a satisfactory atmosphere in which to pursue mathematical research is the presence of a colleague in an allied field, with whom one can share ideas. Mittag-Leffler, as a fellow student of Weierstrass, was an ideal discussion partner. He was always willing to act as a sounding board for Kovalevskaia's inspirations, and she in turn was stimulated by his new ideas.

Sofia kept in touch with mathematical developments in other countries through periodic visits and through her correspondence with such well-known European mathematicians as Charles Hermite, Emile Picard, Leopold Kronecker, and Carl Runge, as well as her Russian colleagues P. L. Chebyshev, A. V. Vasilev, D. F. Selivanov, and others. In fact, the correspondence with Kronecker was extensive enough that for a while, Hermite thought that Kovalevskaia had been Kronecker's student, much to Weierstrass' indignation.[18]

Soon after she arrived in Stockholm, Kovalevskaia began serious research on the classical problem of the rotation of a solid body about a fixed point, which had fascinated her almost from the start of her mathematical career.[19] In a letter to Maria

[17]29 April 1884 letter to Mittag-Leffler, published in Leffler, "Sonya Kovalevsky," p. 212.

[18]See diary entries for 1884, *VIP 1951*, pp. 177–79; letters to Kovalevskaia in the IML; *Briefe/Pis'ma*, p. 108ff.

[19]Kovalevskaia, letter to Mittag-Leffler in Polubarinova-Kochina, "Iz perepiski S. V. Kovalevskoi," *Uspekhi matematicheskikh nauk*, 7, No. 4(50) (1953), p. 107; Clara Weierstrass, 3 March 1882 letter to Kovalevskaia, IML. Examples of motion of the type Kovalevskaia was considering are pendula,

Jankowska-Mendelson, she set the time it would take her to resolve this problem as the limit of her stay in Sweden:

> The new mathematical work I recently began intensely interests me now, and I would hate to die without discovering what I am looking for. If I succeed in solving the problem on which I am now working, my name will be listed among those of the most prominent mathematicians.
>
> According to my calculations, I need another five years to get good results. But I hope that in five years there will be more than one woman capable of taking my place here, and I can devote myself then to other urges of my gypsy nature.[20]

Curiously, it turned out that Kovalevskaia was right about how long it would take her to finish her work. She submitted her paper on the revolution of a solid body about a fixed point to the French Academy of Sciences in December 1888.

Kovalevskaia was mistaken, however, in her prediction of how far the education of women would progress in five years. There would be no women capable of taking her place in Stockholm. Furthermore, there would be no opening for her in universities in Russia or France. In fact, it would be 1906—well into the first decade of the twentieth century—before Maria Sklodowska Curie would occupy a university chair in Europe.

Kovalevskaia was also premature in her confidence that she would be able to stay in Stockholm for the five years she needed to complete her master work. Her appointment was only for a year, at which time it would be reviewed, and there was the antagonism of the traditionalists to be taken into account. In fact, Mittag-Leffler and his colleagues had a difficult battle to get Sofia officially appointed for a five-year term.

Part of the problem was hostility within the university itself. In the ten months since Kovalevskaia's original appointment, her enemies had consolidated their forces. Now, they made it clear that Sofia would not be promoted to a regular position if they could prevent it.

---

gyroscopes, and tops. The problem will be discussed further in Chapter XIII.

[20]19 January 1884 letter to Jankowska-Mendelson in Mendelson, p. 152.

Opposition was directed not only against Kovalevskaia as a woman, but also against her as a disciple of Weierstrass, and as a foreigner. Scandinavian scientists, like their Russian counterparts, were suspicious of "the German tendency" in mathematics, and distrusted people who had been educated abroad. They thought that one former student of Weierstrass—Mittag-Leffler—was one too many, particularly as Gösta was so astute politically and already powerful at the university.[21]

In the end, Kovalevskaia's three main supporters—Mittag-Leffler, the geographer-explorer Eric Nordenskjöld, and the astronomer Hugo Gyldén—consented to make a trade. They would withdraw their opposition to the promotion of two protégés of the Uppsala school from *privat-docent* to ordinary professor, which was the equivalent of an American full professor. In return, Kovalevskaia would be appointed to a five-year "extraordinary" professorship, which was the approximate analog of an assistant professorship. "You see how expensive I am," Sofia remarked humorously to her brother-in-law, "I went for two 'ordinaries'!"[22]

Official notification of Kovalevskaia's new position came through on June 28, 1884. As expected, Sofia had been appointed to an extraordinary professorship, which Mittag-Leffler hoped to make permanent after five years. Her salary would be 4000 Swedish kronor a year, and would be contributed half from the university's general funds, and half from subscriptions.[23]

This sum would be enough for Sofia to live on if she were careful, but it was by no means an extravagant amount. In fact, since Sofia was still paying off Vladimir's debts, and sending money to her wastrel brother Fedia and her daughter Fufa, she

---

[21]Mittag-Leffler, July 1884 letter to Weierstrass, IML.

[22]May 1884 letter to Aleksander, *VIP 1951*, p. 281. In writing to Aleksander in May, Sofia was premature in her joy. Mittag-Leffler did not consider all enemies neutralized until the end of June. It is possible, therefore, that Shtraikh misdated this letter.

[23]28 June 1884 official notification, and subscription list for Kovalevskaia's salary, IML. Mittag-Leffler and the astronomer Hugo Gyldén contributed 100 kronor apiece, as did several women who wished to be Kovalevskaia's students.

was always short of funds. She translated, and periodically pawned or sold things to make ends meet.[24]

Kovalevskaia spent the summer holidays of 1884 in Russia and Berlin. She tried to obtain access to lectures at the university in Berlin, but was unsuccessful, in spite of Weierstrass' intervention. The rector of the university was a "decided opponent of women's rights," as Sofia wrote to Mittag-Leffler, and would not extend to her, as a woman, the courtesy usually given to visiting professors of other institutions. As Anna Carlotta Leffler ironically noted, "the University of Stockholm had already appointed Fru [Madame] Kovalevsky professor, while in Germany, it was still impossible for her, as a woman, to attend even lectures." She was, however, invited to attend the annual meeting of the Berlin Academy of Sciences. The next day the newspapers mentioned that "in the auditorium was Frau Kovalevskaia, professor of mathematics at Stockholm."[25]

Before Kovalevskaia went to Berlin, she traveled to Moscow for two months to be with her daughter Fufa and Iulia Lermontova. She decided that she would not bring Fufa back with her to Stockholm, because she had not yet settled in completely. Her decision deliberately defied the well-meaning, gossipy Swedish matrons who had advised her to bring Fufa back with her so that she would attain full respectability.

Sofia, however, was not concerned with what the Swedish ladies thought. As she wrote to Theresa Gyldén, the wife of her astronomer colleague and one of her friends in Sweden:

> I have to admit that in the resolution of such an important question [Fufa's welfare] I could not care less about 'what people will say.' I'm fully agreeable to bowing to the opinion of Stockholm society in all the trivialities of life. In my dress and style of life and choice of acquaintances and such I meticulously avoid all that could offend the most severe judge—usually female. But when the subject under discussion is such an impor-

[24]See Kovalevskaia, 29 April 1884 letter to Mittag-Leffler, and various documents, scraps of paper, account books, and pawn tickets in the IML.

[25]1 July 1884 letter to Mittag-Leffler, published in Leffler, "Sonya Kovalevsky," p. 213; 14 July 1884 letter to Aleksander, *VIP 1951*, p. 286.

tant one as the welfare of my daughter, then I must behave fully in accord with my own judgment.[26]

Kovalevskaia was having to worry more about gossip as her fame increased. Her appointment to the professorship in Stockholm, even though the position was as yet only a junior one, had led to a rash of newspaper and journal articles about her. Mittag-Leffler was approached by several reporters with requests for photographs and details of Kovalevskaia's life. The German *Illustrated Times*, the Swedish *Illustrated World*, and the British *London Illustrated News* all ran biographies of her.[27]

Sofia was pleased with this recognition, which reached great proportions even in conservative Prussia. The Minister of Education von Gossler voiced a wish to meet her, and as a result of this interest the rector of the University of Berlin was forced to reconsider his decision to ban her from lectures. That winter, Kovalevskaia was given permission to attend all lectures in any university in Prussia. She was the only woman to have that honor.[28]

Kovalevskaia made the most of her months away from the stiff society of Stockholm. She relaxed on Lermontova's estate outside of Moscow, and visited old friends in St. Petersburg. In Berlin, too, she was able to amuse herself. She and her mathematician friends went to the theater and opera, and picnicked outside the city. Nor did her mathematics suffer. Berlin was the perfect research environment. There were always several specialists visiting Weierstrass, and the atmosphere among them was cheerful and stimulating.

Kovalevskaia had several good friends in Berlin. One of them was the physicist Gustav Hansemann, with whom she kept up

[26]May 1884 letter in *VIP 1951*, pp. 283–84.

[27]14 July 1884 letter to Aleksander, *VIP 1951*, p. 286; 1 July 1884 letter to Mittag-Leffler, IML. In this letter, Sofia says she was born in 1853. There is a box containing hundreds of newspaper and journal clippings on Kovalevskaia in the IML. Mittag-Leffler employed a service which collected articles until 1917.

[28]14 July 1884 letter to Aleksander, *VIP 1951*, p. 286; 24 December 1884 document, Kovalevskaia papers, IML.

Sofia Kovalevskaia

a warm correspondence until her death.[29] Sofia confided to
Hansemann her distress at having to return to Sweden at the
end of the summer: "I must confess, I have about the same
spirit as a child who must leave home and return to school,"
she wrote. She appreciated the honor of having a position in
the university, and knew that her Swedish acquaintances
wished her well. But she was sad to have to part with her
friends and colleagues until Christmas vacation.[30]

In late August 1884, upon her return to Sweden, Kovalevskaia went out to the countryside to finish writing an an-
nouncement of the results of her memoir "On the Refraction of
Light in a Crystalline Medium" for publication in the *Comptes
Rendus* of the French Academy of Sciences. Mittag-Leffler and
the German mathematician Carl Runge joined her at the retreat
she had chosen, where they talked about mathematics and
prepared their lectures for the coming year.[31]

Kovalevskaia was serious about her teaching responsibilities,
always planning out her lectures in advance. Her style was
terse and uncompromising, with no attempts at humor or
liveliness. One witness to her lectures, the Russian as-
tronomer-topographer V. V. Vitkovskii, said she would move
briskly to the board, deliver her talk clearly, and then leave the
room swiftly, almost without acknowledging her audience.[32]

This abruptness was no reflection on Kovalevskaia's feelings
about her students. Possibly, she felt that as a woman, especially an attractive thirty-four-year-old, she needed to preserve
a grave exterior while in the classroom. When she was out of
class, she was warm and friendly to her students. She admired
their capabilities, and boasted to her brother-in-law about
them: "I cannot praise my auditors enough. Truly, we get the
best ones—the worst go to Uppsala."[33]

[29]Letters from Kovalevskaia to Hansemann are published in Marie von
Bunsen, "Sonja Kowalevsky," *Illustrierte Deutsche Monatshefte*, No. 82 (1897),
p. 223ff.; Hansemann's letters to Kovalevskaia are in the IML.

[30]3 July 1884 letter to Hansemann in Bunsen, p. 223.

[31]Leffler, "Sonya Kovalevsky," p. 215; Kovalevskaia, 7 September 1884
letter to Mittag-Leffler, IML.

[32]V. V. Vitkovskii, *Perezhitoe* (Leningrad: n. p., 1927), pp. 124–25.

[33]Kovalevskaia, 21 February 1884 letter to Aleksander in Shtraikh, *Sestry*, p.

According to Anna Carlotta Leffler, Kovalevskaia had changed almost beyond recognition during the summer away from Stockholm. She had discarded her mourning outfit, which did not suit her coloring, and was ready to be the instigator of any party or scheme of pleasure that autumn. She took up skating, horseback riding, and embroidery—all without notable success, but with enthusiasm. She was presented at the Swedish court, and went to balls and parties in high society.[34]

But Sofia soon became bored with the Swedish social scene. Anna Carlotta recalled with some hurt that Sofia "said she knew everyone by heart, and longed for fresh stimulus for her intelligence." Kovalevskaia began to talk in a way that insulted her Swedish friends: "The road from Stockholm to Malmö [from whence one can take a boat to Germany] is the most beautiful line I have ever seen; but the road from Malmö to Stockholm is the ugliest, dullest, and most tiresome I have ever known."[35]

Kovalevskaia's visits to Russia were not always enjoyable either. Her sister Aniuta had fallen ill with what seems to have been heart problems and cancer, and would ail until her premature death in 1887. Her brother Fedia had proved to be so incompetent at financial dealings that Sofia had to contribute resources she could ill afford to help him out of his difficulties.

Her own affairs in St. Petersburg were also in disarray. Sofia had eventually managed to save something from the 1880 bankruptcy. She still had the mortgaged baths, an apartment building she and Fedia had inherited from their aunts, and publishing rights to a multi-volume book on birds and other animals by Jacob Brehm. She put these dealings in the hands of her friend Sergei Lamanskii.

---

315. Kovalevskaia was impressed enough by one of her woman students, Julia Kjellberg, to write recommending her to Weierstrass and Frobenius. (13 September 1884 letter from Weierstrass to Kovalevskaia in *Briefe/Pis'ma*, p. 115.)

[34]Leffler, "Sonya Kovalevsky," pp. 216–17. At her request, Kovalevskaia was presented to the Swedish king by the Russian Minister in Stockholm, N. Shishkin. (26 October 1884 letter from Baron Hochschild to Kovalevskaia, IML.)

[35]*Ibid.*, p. 217.

Lamanskii was willing to be of service. He had always admired Sofia's perseverance and mathematical skill. But he was inexperienced in financial matters, and as gullible as Sofia, Vladimir, and Fedia when it came to falling for the blandishments of plausible scoundrels. Lamanskii, too, often made the wrong decision.[36]

In the late spring of 1885, Hjalmar Holmgren, the professor who taught mechanics at the university and technical school in Stockholm, became ill. It was proposed that Kovalevskaia take over his lectures until he recovered, a task which she was willing to undertake. The idea ran into some opposition at the administrative level, but it was decided that Sofia could fill the position temporarily.[37]

There was some hope in Kovalevskaia's mind that the university would appoint her to fill the ailing Holmgren's position permanently. When Holmgren died in early autumn of 1885, the question came before the faculty, but Sofia's candidacy was rejected. Forces hostile to Mittag-Leffler and his allies had become organized. Professors even voiced the opinion that they would rather see the Chair of Mechanics abolished altogether than have it occupied by a "revolutionary" woman like Kovalevskaia.[38]

It is interesting that most of the opposition to Kovalevskaia within the Stockholm University faculty was apparently not because she was a woman. The university clearly supported the education of women, and even its most conservative faculty members were silent, at least in public, on the question of Sofia's sex. Instead, controversy focused more on her nationality, her place as a prominent member of Weierstrass' mathematical school, and especially her politics.

In fact, even Sofia's position as a temporary replacement for Holmgren was put into jeopardy that autumn when her friend

[36]10 October 1884 letter to S. A. Lamanskii, *VIP 1951*, pp. 288–89; Lamanskii's letters to Kovalevskaia, IML; [29 April 1884] letter to Mittag-Leffler from Kovalevskaia, IML.

[37]3 June 1885 letter from Kovalevskaia to Mittag-Leffler, published in Leffler, "Sonya Kovalevsky," p. 225.

[38]Kovalevskaia's diary entries *VIP 1951*, pp. 180–81.

the German socialist Georg Vollmar came to visit Stockholm. Vollmar was a well-known radical, and had been arrested several times in the past few years. The sight of Kovalevskaia with Vollmar aroused indignation on the part of the more conservative contributors to Stockholm University's finances. They announced that they had no intention of paying a nihilist the three thousand extra kronor pledged for her salary as Holmgren's substitute. Only Mittag-Leffler's political adroitness saved the situation.[39]

Even the tactical skill of Mittag-Leffler was not enough to aid Kovalevskaia later that academic year, however. Gösta had been planning to make Sofia's position more secure by having her elected to the Swedish Academy of Sciences. He wrote a petition arguing that "man" should be changed to "person" in the membership rules of the Academy. He obtained letters of recommendation for Sofia from Charles Hermite and Lazarus Fuchs, and Hermite collected signatures of support from other prominent mathematicians.[40]

Mittag-Leffler made his case for Kovalevskaia's election coherently and well. But eloquence and logic in this instance were not enough to sway the conservative members of the Academy of Sciences. Academy members were largely drawn from the established universities of Uppsala and Lund, and for them, Sofia's sex was a consideration. Professor Wittrock, a botanist, argued that a woman by her nature could never attain the standards necessary for an Academy post.

Wittrock and his anti-woman allies were unwittingly aided by the mathematician Leopold Kronecker. Kronecker was, in general, well-disposed toward Kovalevskaia. He praised her work to his colleagues, and later wrote a glowing obituary of her, calling her "the rarest investigator."[41] But unfortunately,

---

[39]Mittag-Leffler, 26 November 1885 letter to Weierstrass, IML. Vollmar, incidentally, met Kovalevskaia's student Julia Kjellberg and married her that year.

[40]Charles Hermite, 22 March and 13 May 1886 letters to Mittag-Leffler; Mittag-Leffler, "Motiver" and other documents of the Swedish Academy of Sciences concerning Sofia's nomination, dated 13 January 1886 and 5 December 1885, IML.

[41]L. Kronecker, "Sophie von Kowalevsky," *Crelle's Journal*, 108 (1891), p. 88.

Kronecker was inspired by one of his periodic disputes with Weierstrass to write an unsolicited letter opposing Kovalevskaia's election.

As if all this academic maneuvering were not sufficient to ensure the defeat of Mittag-Leffler's motion, a series of smear letters were sent to the Academy as well. Kovalevskaia is a revolutionary, the letters ran. She consorts publicly with criminal insurgents (i.e., Volmar) and openly shares their political views. Such a person has no place in our Academy of Sciences, whatever her scientific qualifications for the post might be.[42]

Mittag-Leffler, Hermite and Weierstrass were distressed by the hostility against Kovalevskaia, and worried about the effect of rumors on Sofia's position at the university. Unfortunately, Weierstrass mused sadly, she does leave herself open to charges of radicalism. He found himself wishing Sofia would not be so obvious in her sympathies to the socialists. Her political convictions made matters even more difficult for her supporters than they would otherwise be.[43]

The affair of the Swedish Academy of Sciences was neither the first nor the last time Kovalevskaia's political convictions were used against her by her opponents. Sofia knew that she could have disarmed some of her critics by renouncing her radical sympathies, but it would never have entered her head to do so. As she had told Theresa Gyldén when the latter reproached her about Fufa's upbringing, on important matters Sofia followed only the dictates of her own conscience.

This firmness on Kovalevskaia's part would often cause problems, and would repeatedly exasperate Mittag-Leffler and Weierstrass. But they, like virtually all mathematicians, felt that Kovalevskaia's politics and lifestyle were her own affair. If university officials and professors in other fields thought differently, then it was the job of Mittag-Leffler and other mathematicians to protect Kovalevskaia's interests. In the years to come, there would be several opportunities for the mathematical community to band together in support of their first female colleague.

---

[42]Mittag-Leffler, 7 April and 4 June 1886 letters to Weierstrass, IML.

[43]Weierstrass, 7 May 1886 letter to Mittag-Leffler, IML.

eleven

A WOMAN IN THE EUROPEAN MATHEMATICAL COMMUNITY

Unlike Mittag-Leffler, Sofia Kovalevskaia had little tactical sense. Nevertheless, in 1885–86 she seems to have realized that she needed to appease her opponents at the Swedish Academy of Sciences and at Stockholm University. The retention of her position was important to Kovalevskaia for political as well as mathematical and personal reasons: she wanted to establish a precedent for women as university teachers.

Since Sofia had no intention of changing her political beliefs or her lifestyle, she came to the conclusion that perhaps she had best make some symbolic gesture to mollify her Swedish critics. She decided to apply for Swedish citizenship, thinking that in this way she would blur her connection with the dreaded Russian revolutionary movement in the eyes of Stockholm conservatives.

After making inquiries, Sofia learned that first she had to reject her Russian citizenship. This turned out to be more complicated than she had envisioned, since she would have to discharge all her Russian debts. This she could not do, so she abandoned the project. She had not really wanted to change her passport, since she was first of all a Russian. Moreover, it was soon apparent, even to someone as naive as Sofia, that opposition to her, and to Mittag-Leffler as well, was too serious to be silenced by such a token gesture.[1]

[1]6 July and 9 October 1885, 16 May 1886 documents from the Swedish and

The furor over Sofia's nomination to Holmgren's chair of mechanics and membership in the Academy of Sciences had shaken Mittag-Leffler's confidence. He knew that what he called "the cabal" was directed not only against Kovalevskaia herself, but also against him and Weierstrass. Weierstrass was head of the mathematical school which was thought by many to indulge in "mathematical analysis for analysis' sake," that is, to have abandoned the concrete practical problems that concerned other analysts. Sofia and Gösta were prominent members of this school of analysis. Mittag-Leffler had recently founded a journal, *Acta Mathematica*, partly to express this tendency in mathematics.

Soon after Kovalevskaia came to Stockholm, Gösta asked her to become an editor of the journal. Sofia accepted with pleasure, thereby becoming the first woman in the world to be on the editorial board of a major scientific journal. Sofia took her job seriously, propagandized for *Acta* all over Europe, and unsuccessfully tried to get financial support for the journal from the Russian Academy of Sciences. With Mittag-Leffler, she soon became identified in European mathematicians' minds as one of the main sponsors of *Acta Mathematica*.[2]

Consequently, when people wished to attack Weierstrass and his school, they naturally attacked Kovalevskaia, Mittag-Leffler, and *Acta Mathematica*. Sofia's, Gösta's, and Weierstrass' correspondence is filled with references to such attacks and feuds. But, because they felt that *Acta* was more important than petty mathematical jealousies, Sofia and Gösta tried to rise above these quarrels. They solicited manuscripts from all good mathematicians, even from such critics of Weierstrass' method as Kronecker.

As an editor of *Acta Mathematica*, Kovalevskaia conducted a constant correspondence with European and Russian mathematicians. Weierstrass consulted her about the form of his articles, Hermite consented to help her with French papers sub-

---

Russian Ministers in Stockholm and St. Petersburg; Kovalevskaia, 24 May 1885 letter to Mittag-Leffler; all in the Kovalevskaia papers of the IML.

[2]Interestingly enough, some of the present editors of *Acta Mathematica* seem to prefer to minimize the journal's connection with Kovalevskaia. See my article on Kovalevskaia, to appear in *The Mathematical Intelligencer*, for details.

mitted to *Acta*, Kronecker enlisted her aid in the mathematical feuds for which he was famous.[3]

For Russian mathematicians, Kovalevskaia served as a conduit for making their ideas known in Western Europe. P. L. Chebyshev wrote gratefully of her efforts in that direction, as did A. V. Vasilev, S. N. Tsvet, A. P. Starkov, and others. Apparently, Russian mathematicians of even mediocre attainments felt that she would give their papers a fair review, and so they sent their researches to her rather than to Mittag-Leffler.

At the same time Kovalevskaia was involved in editing *Acta Mathematica*, she was also working on the revolution problem. Her research received an added impetus when Hermite encouraged her. He implied that the French Academy of Sciences was thinking of offering a prize for which her work would be suitable.

With her editorial obligations, teaching, and research, Kovalevskaia found herself busy indeed. Although she was only living in furnished rooms, there were still domestic chores to be attended to—food to be bought, clothes to be kept in order, and so on—and she resented the time they took. With characteristic humor and irony she complained to her friend Hansemann: "All these stupid but unpostponable practical affairs are a serious test of my patience, and I begin to understand why men treasure good, practical housewives so highly. Were I a man, I'd choose myself a beautiful little housewife who'd free me from all this."[4]

Coincidentally, around this time Kovalevskaia invited Iulia Lermontova to come live with her in Stockholm. They could raise Fufa together, she said. Iulia refused, possibly because she knew her old friend's views on household concerns, and did not choose to become Sofia's "housewife."[5]

As soon as the summer holidays began, Kovalevskaia traveled to France, where she again stayed with Maria Jan-

[3]See letters in *Briefe/Pis'ma*, p. 122ff.; letters from Hermite, Kronecker, Chebyshev, Duhem, Vasilev, and many others in the Kovalevskaia papers of the IML.

[4]September 1885 letter in Bunsen, p. 226.

[5]Kovalevskaia, September 1885 letter to Lermontova, *VIP 1951*, pp. 290–92.

kowska-Mendelson. She met constantly with French mathematicians, and discussed with them her work on the rotation problem. Several of her most prominent French colleagues—Hermite, Joseph Bertrand, Gaston Darboux, and Emile Picard—assured her that her work would win the Academy prize; Bertrand, the secretary of the Academy, told her the deadline would be mid 1888.

Kovalevskaia was pleased. Her work was going wonderfully, and her Parisian professional acquaintances were more cordial than ever. In addition, their wives seem suddenly to have discovered that Sofia was harmless. That summer, she received invitations from the wives of Poincaré, Lippmann, Bertrand, and even the formerly hostile Madame Hermite.[6]

Kovalevskaia returned to Scandinavia for a natural scientists' congress, and spent the month afterward pleasantly in a tour of Sweden with Anna Carlotta and the Mittag-Lefflers. On this tour she visited a peasant school and was impressed with the intelligence and diligence of the students. She later wrote an article about it for the Russian journal *Severnyi vestnik* (Northern Messenger).[7]

In the autumn of 1886, in spite of her absorption in the rotation problem, Kovalevskaia decided to bring her daughter back to Stockholm with her. She set up a household of sorts, with cheap furniture bought in Stockholm and old pieces from Palibino. Anna Carlotta described the Palibino furniture with amusement not unmixed with horror. Originally elegant, the pieces were now worn and tawdry-looking—ugly, but in tune with Sofia's notions of housekeeping.[8]

Kovalevskaia had never been one to care much about appearances, even when she had the money and time to arrange her household neatly. Now, when she was so busy, was short of funds, and had Fufa to take care of, she paid no attention to her surroundings. She covered up the holes in the Palibino furniture with antimacassars, tossed cloth over the worn spots, and

[6]Kovalevskaia [18 June] and 26 June 1886 letters to Mittag-Leffler, IML.

[7]See "Tri dnia v krest'ianskom universitete v Shvetsii," *VP 1974*, pp. 244–67.

[8]Leffler, "Sonya Kovalevsky," p. 237.

considered herself ready to entertain whomever came to visit.[9]

Much to Kovalevskaia's surprise and amusement, her Swedish acquaintances in no way felt that Sofia's decorating was even passable, let alone adequate for the reception of visitors. They pointed out the holes, the scratches, the missing leg on one cabinet, and could not accept that Sofia intended to leave everything as it was. To be sure, she entertained herself and them with grandiose redecorating schemes, but everyone knew she was only fantasizing.

Anna Carlotta thought this unconcern proved that Kovalevskaia had no intention of remaining in Sweden, and considered Stockholm merely a "half-way house" on her way to an appointment in some large European capital. This reasoning was incorrect, although the conclusion—that Sofia would leave Stockholm if she were ever offered an analogous position in Russia, France, or Germany—was a true one.

Kovalevskaia simply had no interest in those domestic talents generally considered to be the "true sphere" of women. She could not be bothered dressing up her living quarters to appeal to others' standards of neatness and style. She once chided Anna Carlotta for excessive preoccupation with these standards:

> Anna Carlotta! Yesterday evening I had a pleasant proof that the critics are right who maintain that you have eyes for the bad and ugly, but not for the good and beautiful. Each stain, each scratch, on one of my old venerable chairs, even if hidden by ten antimacassars, is very certain to be discovered and denounced by you. But my really lovely new rocking chair *cushion*, which was in evidence the whole evening and which endeavored to draw your attention to itself, was not honored by you with even a single glance.[10]

Kovalevskaia's slapdash housekeeping in no way kept society away from her, though. She was such a good conversationalist and such a celebrity that the Swedes forced themselves to overlook the inadequacies of her household. They

[9]Litvinova, pp. 46, 69–70; S. Vl. Kovalevskaia, "Vospominaniia," *Pamiati*, pp. 145–46.

[10]Note in Leffler, "Sonya Kovalevsky," p. 238.

criticized her, and told her that Fufa would never learn domestic virtues. But Sofia's home was a gathering place for Swedish society nonetheless.

Sofia had hardly had time to settle herself and Fufa into their new apartment when she was called back to Russia to care for her sister Aniuta. Her sister had been ailing for some time, but in late autumn 1886 her illness became more serious. Aniuta had cancer, was in great pain, and wanted Sofia near her.

Kovalevskaia left Fufa with Anna Carlotta Leffler and Theresa Gyldén, and stayed two months by her sister's bedside. It was painful for Sofia to watch Aniuta waste away. She wrote to Theresa Gyldén: "No, there is nothing more horrible than these diseases that sneak up on one so quietly . . . It is terrible to see a person suffering, and be absolutely unable to help in any way."[11]

The long hours spent at Aniuta's side turned Sofia's thoughts toward the girls' youth, when everything had seemed so golden, and the future so bright. She mused on Aniuta's life—her political development, her participation in the Paris Commune, her literary talent. Aniuta had had such promise, at least in her sister's eyes, and all that promise had come to this: mortal illness, partial separation from a husband notorious for his extra-marital affairs, and a handful of little-known children's stories and novellas.[12]

Kovalevskaia wanted somehow to express her frustration at the way events had turned out. She evolved the idea for two plays: one which would chronicle how life was, and one which would describe how it might have been. They were to be titled *The Struggle For Happiness*. Along with *Memories of Childhood*, they would constitute Sofia's memorial to her sister.

At this time in her life, Kovalevskaia had not yet discovered that she was talented in writing as well as mathematics. She had written only a few attempts at poetry, the group of theater

[11]December 1886 letter in *VIP 1951*, pp. 296–97.

[12]That Jaclard was an unsatisfactory husband is clear from Litvinova and her pseudonym E. El', *passim*; Mendelson, p. 147; and especially Shtraikh, *Sestry*, p. 294ff. In the twenties, Shtraikh collected reminiscences of people who remembered Jaclard. They are almost unanimous on his lack of domestic virtues and blatant infidelity to Aniuta.

reviews and popular science articles in *Novoe vremia*, and a reminiscence of her acquaintance with the writer George Eliot, published in the Swedish press in 1885.[13] She convinced Anna Carlotta Leffler to do the actual writing, following a plot of Sofia's making.

Anna Carlotta was working on a new novel, which was to chronicle the lives of unmarried women. But as usual when Sofia became enthusiastic, the Swedish woman was unable to withstand her. Anna Carlotta wrote: "So great was her influence upon me, so great her power of persuasion, that I forsook my own child in order to adopt hers."[14]

The two women worked so rapidly that their dramas were finished by mid-March 1887 and ready to be read to friends. To Sofia's and Anna Carlotta's astonishment, their acquaintances were not impressed. Leffler mused that she and Sofia had seen their plays as they might have been, rather than as they were. Their friends pointed out the logical gaps, thinly drawn characters and awkward dialogue. The two women set about the much less interesting process of editing and revising their brainchild.

For Kovalevskaia, this was not the time to be writing plays, because the deadline had already been fixed for the Prix Bordin competition. As Mittag-Leffler pointed out to her, she needed to devote herself to her work on the rotation problem. It was a difficult time, however, because Sofia had recurring calls to Aniuta's sickbed as well as revisions on *The Struggle For Happiness* to distract her attention.

The summer of 1887, Kovalevskaia was summoned to face a crisis in Russia. Jaclard, who had returned to Russia to bring the ailing Aniuta back to Paris, was given three days to leave the country or face arrest. Victor had been in regular contact with the Polish Proletariat Party, of which Maria Jankowska-Mendelson and her husband Stanislaw Mendelson were mem-

[13]See "Vospominaniia o Dzhorzhe Elliote [sic]," *VP 1974*, pp. 230–44. Kovalevskaia had originally written the article in Russian, and her friends translated it for publication in *Stockholms Dagbladet*.

[14]Leffler, "Sonya Kovalevsky," p. 241.

bers.[15] Moreover, in his role as Russian correspondent for several French newspapers, Jaclard had written approvingly of the plot by a group including A. I. Ulianov (Lenin's older brother) to assassinate Alexander III.

Aniuta and Victor were both frantic. There was no way they could leave Russia within the specified time. Aniuta needed to be carried on a stretcher, and the train journey would certainly kill her if she did not have frequent rests. She appealed to Dostoevskii's widow, with whom she was acquainted. Anna Dostoevskaia in turn appealed to the wife of the influential reactionary, K. P. Pobedonostsev, who wrote to the head of police, and obtained an extension of ten days.[16]

Even this extension was not enough, because Aniuta suddenly became worse and could not be moved. Jaclard was forced to leave without her, and Aniuta was left in Sofia's care until this most recent attack had passed.

Kovalevskaia needed some sort of occupation to distract herself while she sat by her unconscious sister's bedside. Whenever she could, she took up her mathematical research. At times, concentration on mathematics relieved her anxieties. She told Anna Carlotta: "I think out my mathematical problem, and muse deeply upon Poincaré's treatises, which are full of genius. . . . At such moments mathematics are a relief. It is such a comfort to feel that there is another world outside one's self."[17]

After a time Aniuta recovered enough to be sent to Paris in the care of friends. She was still dangerously ill, but her disease seemed to be in remission. Kovalevskaia remained in Russia for a few weeks to recuperate after nursing Aniuta. She renewed old acquaintances, made new ones, and visited the mathematics department of the Higher Women's Courses. She spent some of her time with Maria Bokova-Sechenova and Ivan Se-

---

[15]For a history of this party see Norman Naimark, *The History of the "Proletariat": The Emergence of Marxism in the Kingdom of Poland, 1870–1887* (New York: East European Quarterly, 1979).

[16]See letters of Anna Dostoevskaia, Police Chief P. N. Durnovo (later Minister of the Interior), and Jaclard, in Shtraikh, *Sestry*, pp. 300–02.

[17]Kovalevskaia, undated [summer 1887] letter to Leffler in Leffler, "Sonya Kovalevsky," p. 253.

chenov. In their loving, understanding company, she regained her vitality and composure.[18]

Kovalevskaia returned to Sweden determined to make up for her mathematically unproductive spring and summer. She rented rooms on one of the islands not far from Stockholm, and worked steadily on the rotation problem. She was beginning to feel confident that she could submit a finished paper in time for the Prix Bordin deadline. She only needed to work efficiently during the next year.

To some extent, events conspired against Kovalevskaia's desire to devote herself to her own researches. She and Anna Carlotta continued to make revisions on *The Struggle For Happiness*, although their initial enthusiasm for the project had waned. Sofia had her lectures at the university to deliver, and her work for *Acta Mathematica* as well. On top of all this, word came that Aniuta had died suddenly in Paris after an operation which had been described earlier as a complete success.

Sofia was stunned by Aniuta's death. It had been expected and, in light of Aniuta's constant pain, Sofia could not help but find death a release for her sister. But Sofia had assumed that there would be time enough at the end for her to journey to Aniuta's deathbed.

Moreover, Sofia regretted that Aniuta had not been in Russia when she died, among the friends Aniuta considered so much more sincere than her French acquaintances.[19] She must have been particularly upset because Aniuta had been extremely unhappy right before her death. In August 1887 she wrote to Sofia: "And I? I confess completely openly—I curse the day when I decided to travel to Paris. After all, if I must die, then it would be better to die in Russia. You know death does not worry me: I want to die . . ."[20]

Kovalevskaia's sorrow over her sister's death was more subdued, though undoubtedly deeper, than that she had felt at her husband's suicide. She continued with her usual activities, and wore no mourning, since both she and Aniuta detested such

[18]Panteleev, "Pamiati S. V. Kovalevskoi," *Rech'*, 29 January 1916.

[19]Aniuta [August 1887] letter to Sofia, Kovalevskaia papers, IML. Roger Cooke originally found and transcribed the letter.

[20]*Ibid.*

outward show. But her *Memories of Childhood,* which she began soon after Aniuta's death, were a testimonial to the love and gratitude Kovalevskaia felt toward her sister. In these memoirs, she paid the ultimate tribute: she gave Aniuta the center stage, and relegated herself to the role of adoring acolyte.

The contrast between Kovalevskaia's reactions to the deaths of Aniuta and Vladimir was marked. Aniuta had influenced her sister, while Vladimir, in the last analysis, had been a burden. Yet Sofia mourned her husband in an ostentatious manner, while outwardly she did not mourn Aniuta at all. This seeming paradox can perhaps be explained by something Kovalevskaia once wrote to Mittag-Leffler: "I only bemoan and bewail when I am *slightly* unhappy. When I am in great distress, then I am silent. No one can detect my anguish."[21]

In the winter of 1887–88, Kovalevskaia continued to do a certain amount of socializing, even after she learned of her sister's death. Some of this was necessary and, one might say, a part of her job. Stockholm University was a young, private institution which subsisted largely on the contributions of well-to-do citizens. Sofia, Gösta, and their colleagues had to entertain for the purpose of fundraising. They needed to circulate in respectable bourgeois society, charm influential people, and convince them to lend their support to the new university.

The round of parties and visits was sometimes tedious, especially since several merchant bankers and financiers (including the famous Alfred Nobel and his brothers) showed some inclination to court Sofia. Sofia does not seem to have been particularly flattered by their pursuit of her. She mentions the Nobels ironically in her letters to Mittag-Leffler, but was cordial to them for the sake of the university.

But Sofia sometimes met interesting people as well. In 1887 she was introduced to the arctic explorer Frithiof Nansen, to whom she was attracted. He reciprocated her liking, and they might have fallen in love were it not for two circumstances. Nansen, it turned out, was already engaged. And Kovalevskaia

[21]Kovalevskaia, 3 June 1885 letter to Mittag-Leffler, IML.

had just met a Russian who fascinated her even more than did Nansen—Maksim Maksimovich Kovalevskii.[22]

Kovalevskii was a well-known figure among Russian and European intellectuals.[23] He had been a professor of law at Moscow University until he and several other professors were dismissed in 1887 during a wave of reaction which followed the discovery of an assassination plot on the tsar. He had been known for his political views, which were liberal (but not radical), and his popularity with students. After his dismissal, he lectured in London, New York, Paris, Stockholm, and other European and American cities. He wrote learned treatises on the development of social institutions in the style of Herbert Spencer, Henry Maine, and John Lewis, and was acquainted with many of the intellectual celebrities of his day.

Maksim Kovalevskii, who was a distant relative of Sofia's late husband, Vladimir Kovalevskii, came to Stockholm to deliver a series of lectures. The lectures were financed through the estate of a young economist, who had set aside a substantial amount of money for the study of "social questions"—political economy, sociology, agricultural economy, labor issues, and so on. Anna Carlotta Leffler and Kovalevskaia were among those named as executors of the estate, and so had some correspondence with Kovalevskii.

When Maksim arrived in Stockholm, he naturally looked up his countrywoman, whom he had once met in Lavrov's apartment in Paris. They became close, Kovalevskii explained, because they were both homesick for Russia: "She was treated with respect and even enthusiasm, but without warm intimacy. Always she felt herself a Russian woman, torn from her usual surroundings, living for Russian interests, seeking above all a more sincere discussion of what was happening on this side of the Baltic [that is, in Russia]."[24]

[22]There is a packet of friendly, chatty letters from Nansen in the Kovalevskaia papers of the IML.

[23]For descriptions of Kovalevskii's life and work, see A. P. Kazakov, *Teoriia progressa v russkoi sotsiologii* (Leningrad: Leningradskogo Universiteta, 1969); *M. Kovalevskii 1851–1916. Sbornik statei* (Petrograd: Marks, 1918); I. Ivanovskii, *Maksim Maksimovich Kovalevskii* (Petrograd: Volf, 1916).

[24]"M. M. Kovalevskii o S. V. Kovalevskoi," *VIP 1951*, p. 389. This section

Kovalevskaia, on her part, was enthralled with Maksim, who struck her as the epitome of all things Russian. She spent so much time with him that Mittag-Leffler persuaded him to go to Uppsala for a while, so that Sofia could work in peace. She wrote to Anna Carlotta Leffler that she was glad of this change in plan, because "if stout M— had stayed longer, I do not know how I should have got on with my work."[25]

Kovalevskaia was in the final stages of her research on the revolution of a solid body about a fixed point, and needed only quiet to complete her project. In Maksim's absence, she worked day and night. Within a few weeks, she had finished enough to send a first draft to the French Academy of Sciences, with the request that she be allowed to send in a fuller version in time for the judging.[26]

There was no rest for Kovalevskaia that spring, summer, or fall. She wrote to Weierstrass with her partial solution, and asked for advice. Weierstrass was ill, and could not answer her letter. But meanwhile, Sofia was working so steadily that she arrived at a better solution herself. Her former teacher wrote: "Your letter of the 17th shows me, to my great happiness, that you have come a good piece forward . . ."[27]

Although Kovalevskaia did not know it, there was no reason for her to work so feverishly on perfecting her research on the rotation problem. In fact, if Sofia had been unable to complete her paper on time, in all likelihood Charles Hermite and Joseph Bertrand would have somehow managed to postpone the judging of the Prix Bordin until she was ready. French and German mathematicians had known of Sofia's work since 1885, and there was no doubt in the mind of anyone except Sofia that her work deserved the prize.

---

includes all of Maksim's published reminiscences of Kovalevskaia and excerpts from his unpublished memoirs.

[25]Undated [March 1888] letter in Leffler, "Sonya Kovalevsky," p. 263. Leffler never mentions Maksim by name, because he was still alive.

[26]"M. M. Kovalevskii o S. V. Kovalevskoi," p. 389; Leffler, "Sonya Kovalevsky," p. 265. Entries were submitted anonymously. A saying of some sort headed the entries, and an envelope with the contestant's name inside and the saying outside accompanied the entry.

[27]Weierstrass, 22 May 1888 letter to Kovalevskaia, *Briefe/Pis'ma*, p. 134.

There is even some evidence that mathematicians in the French Academy of Sciences decided to announce a Prix Bordin concourse precisely because Kovalevskaia's work was so good. The prize, the second most prestigious in a long list of Academy awards, was not given every year. Some French mathematicians appear to have felt that Sofia deserved some outstanding recognition for her mathematical research, and fastened on the Prix Bordin as the most appropriate and impressive way of honoring her abilities.[28]

Kovalevskaia had no way of knowing the behind-the-scenes maneuverings of her French colleagues. Gösta Mittag-Leffler and Weierstrass knew, because Hermite had informed them of the plans of the Academy.[29] But they had no intention of telling Sofia. They both knew that she worked better under pressure, and they felt that she had enough distractions without also knowing that her results to date were already sufficient to win the prize. As Gösta complained to Weierstrass, "all the French mathematicians" were waiting for Sofia's memoir, while she, to his "greatest regret" was now working on literature.[30]

As Maksim Kovalevskii writes in his memoirs of Kovalevskaia, the final version of her paper on the rotation problem was written before his eyes. She traveled with him that summer to London, and was in a frenzy of activity. Not only did she work consistently on her research, but she also concentrated on literature and history. She was always interested in politics, and studied seventeenth-century English social movements—the "Diggers" and the "Levelers"—with enthusiasm. She even talked of writing a historical novel about them after she finished her mathematical work.[31]

Kovalevskii admired Sofia's tenacity, although it meant that she had less time to spend with him. In his memoirs he wrote that as friendly as she was to him on their trip to London, her

[28]See Hermite's 1888 letters to Mittag-Leffler, IML.

[29]Charles Hermite, 17 July and 17 October 1888 letters to Mittag-Leffler, IML.

[30]Mittag-Leffler, 15 May 1888 letter to Weierstrass, IML.

[31]"M. M. Kovalevskii o S. V. Kovalevskoi," pp. 389–90.

emotions "did not in the slightest way prevent her from with-drawing into scientific work at any time, and spending whole nights in the resolution of complicated mathematical prob-lems."[32]

In late July 1888, Kovalevskaia left London and Maksim to stay with Weierstrass, his sisters, and some of his other former and present students and their families in the Harz Moun-tains.[33] It was a beautiful place for an informal conference. With Mittag-Leffler, Hurwitz, Hettner, Volterra, and others in atten-dance the mathematical conversations were lively and informa-tive. Sofia could not participate as much as she would have liked, because she was so busy with her paper. But the visit was still refreshing, both physically and mathematically.

Kovalevskaia returned to Stockholm for the fall semester, and continued to tax her strength. Finally finishing her memoir on the rotation problem, she sent it off to the French Academy. It was not a full solution of the problem, but it contained significant new results. Both she and Weierstrass were pleased with the work. Sofia settled down to wait, as patiently as she could under the circumstances, and resumed lectures at the university.

That autumn, Kovalevskaia was irritable and over-excited.[34] She corresponded with Maksim, but letters were no substitute for the long, rambling, philosophical/political/personal conver-sations so dear to the hearts of Russians. Moreover, she was busy. She had had no time to prepare her lectures, and she still had to contend with correspondence for *Acta Mathematica*. Above all, she was nervous about the outcome of the Prix Bor-din competition.

In December 1888, the news came. Fifteen entries had been submitted to the Prix Bordin committee. One of them was con-sidered so original, and so far above the other papers, that the

[32]*Ibid.*, p. 392.

[33]Weierstrass wrote quaintly to Kovalevskaia, advising her to use "Frau von" in front of her name to indicate her nobility and marital status when she wrote for a room. He was convinced that, as a single Russian woman, she would get better service that way. (13 July 1888 letter, *Briefe/Pis'ma*, p. 145.)

[34]S. Vl. Kovalevskaia, "Vospominaniia," *Pamiati*, p. 151.

prize was raised from 3,000 to 5,000 francs. That paper was Sofia Kovalevskaia's.

Kovalevskaia had Hermite and Bertrand to thank for the increased prize money. Earlier in the year, Hermite had written to ask Mittag-Leffler whether it was true that Sofia was rich; such a rumor was circulating among French mathematicians, he said. Mittag-Leffler had replied with the true story of Sofia's straitened financial circumstances. Hermite relayed this information to other Academy members. Since they already had been thinking of raising the prize amount to honor the exceptional merit of Sofia's work, news of her relative poverty decided them in favor of adding another 2,000 francs to the purse.[35]

Kovalevskaia was grateful for the extra money, and ecstatically happy about the Prix Bordin. Five years of research had been fittingly rewarded—with recognition throughout Europe and much-needed cash. Moreover, her personal life was going well. She and Maksim traveled to Paris together, and he accompanied her to the prize-giving ceremony, official Academy of Sciences and Ministry of Education banquets, and dinners given by her Parisian colleagues.

There would be triumphs to follow, in the last two years of Sofia's life. She would be acclaimed as a writer as well as a mathematician, and be feted all over Europe. Her own country would acknowledge her with corresponding membership in the Imperial Academy of Sciences. Her adopted country would give her a lifetime professorship in Stockholm. But her professional accomplishments would not be accompanied by the recognition she most desired—a position in a Russian university. And her life would be cut short in its prime.

[35]Charles Hermite, 17 July and 17 October 1888 letters to Mittag-Leffler, IML. Note that the increase in prize money for Kovalevskaia was already being discussed in July!

twelve

KOVALEVSKAIA'S FINAL YEARS

Kovalevskaia arrived in Paris in mid-December, and received the Prix Bordin of the French Academy of Sciences on December 24, 1888, just weeks before her thirty-ninth birthday. She was only the second woman ever to receive a significant prize from the Academy. In 1816, Sophie Germain won the Grand Prix for her work on the elasticity of metals.

Newspapers throughout Europe acclaimed her achievement, and Sofia was honored at many official banquets and private dinners in the following weeks.[1] Her French colleagues were more cordial than ever. She and Maksim were invited to more celebrations and dinners than they could manage to attend. Maksim had business at his villa in Nice, however, and had to leave Sofia to savor the adulation of Paris alone.

But Kovalevskaia was in no mood to enjoy the praise of her colleagues. Work on the rotation problem had drained her of energy. She wrote to Mittag-Leffler requesting a leave of absence from Stockholm University for reasons of health, and decided to stay in Paris during the spring of 1889. She felt unhappy and unsettled, and was experiencing the sense of anticlimax that often comes after a long period of sustained intellectual effort. With characteristic self-dramatization, she informed Gösta: "I receive so many letters of congratulation, and, by a strong irony of fate, I have never felt so miserable in

---

[1]Kovalevskaia, 12 and 17 January 1889 letters to Mittag-Leffler; Hermite, 19 and 31 December 1888 letters to Mittag-Leffler, all in the IML.

my life."[2] (This melancholy passage, by the way, is followed by a spritely discussion of mathematics.)

Kovalevskaia's depression after completing her work for the Prix Bordin competition was not unusual for her. She had behaved in a similar manner several times before in her life. In fact, her passive state appears reminiscent of those which overcame her after the completion of her doctorate in 1874, and to a lesser extent, after her first provisional appointment in Stockholm. She had worked toward each of these goals in turn—her doctorate, a university position, the Prix Bordin—and achievement left her exhausted and restless. She was ready for the next obstacle, the next challenge, but at the same time she was reluctant to embark on anything new. This is a common feeling among scientists and other creative people, although Sofia sometimes carried it to extreme lengths.

Karl Weierstrass, who knew the defects of Sofia's character as well as anyone, was alarmed by her torpor. From his letters to her, it is clear that he was afraid that she would fall into another five-year hiatus from mathematical studies. He strove to prevent this, pointing out how impressed mathematicians were with her plans for future work, and reminding her that her goal should now be a permanent professorship in Stockholm. He advised her not to take a whole semester off, and even asked Mittag-Leffler to argue with her about it.

Maksim Kovalevskii was also alarmed. Possibly, he feared that Kovalevskaia was reading too much into their relationship, which at this time he saw as one of friendship. Perhaps also he felt that he needed to shock her out of the melancholy into which she had fallen on receipt of the Prix Bordin. In any case, he wrote her a brusque, businesslike letter, advising her to return to Stockholm and resume teaching as soon as possible.[3]

Kovalevskaia remained firm on her need for a leave of absence, and indeed, she was right. She was thirty-nine years old; she was getting to the age where she could not afford to

[2]Kovalevskaia, 12 January 1889 letter to Mittag-Leffler, IML.

[3]Weierstrass, 1 February 1889 letter to Kovalevskaia in *Briefe/Pis'ma*, pp. 146–47; Weierstrass, 2 February 1889 letter to Mittag-Leffler, IML; M. M. Kovalevskii, 2 January 1889 letter to Kovalevskaia, in E. Kannak, "S. Kovalevskaia i M. Kovalevskii," *Novyi zhurnal*, No. 39 (December 1954), p. 204.

neglect her health. Her daughter recalls that Sofia had to be hospitalized at this time, so great was her exhaustion.[4] Although Sofia attended seminars and talked frequently with Hermite, Picard and Poincaré, for the time being she set aside original mathematical research. During the spring of 1889, she slowly recovered from the strain of overwork.

Kovalevskaia took an apartment in Paris, and traveled from there to other parts of Europe. In February, she visited Maksim Kovalevskii at his villa in the south of France. She had a wonderful time, engaging in lively discussions with the Russian émigrés and visitors in the vicinity—the writer P. D. Boborykin, the playwright Anton Chekhov, the journalist V. I. Sobolevskii, and others. In addition, there was a Russian zoological station nearby, so the Kovalevskii villa was visited by a constant stream of Russian and European naturalists; among them were Aleksander Kovalevskii, Ivan Sechenov, Kliment Timiriazev, and Carl Vogt.[5]

While visiting Maksim's villa in Nice, Kovalevskaia happened to recount several stories of her childhood that she was writing up for her own amusement and possible publication. Kovalevskii's literary friends were so enthusiastic about them that she was encouraged to work on them more seriously. During that spring, summer and fall, whenever she was in the mood for writing she added to her memoirs.[6]

Kovalevskaia returned to Paris for the international exhibition to be held there in the spring of 1889. She had a lively social life with her émigré friends—Lavrov, Jankowska-Mendelson, and others—and active mathematical contact with her French colleagues. She wrote to her friend Hansemann: "During this time [in Nice] I did absolutely no mathematics . . . [but now] I am completely well and fit for work. As my leave lasts another two months, I have returned to Paris to carry on my researches in the area of mechanics."[7]

During the spring and summer, Kovalevskaia attended sev-

[4]S. Vl. Kovalevskaia, "Vospominaniia," *Pamiati*, p. 151.

[5]"M. M. Kovalevskii o S. V. Kovalevskoi," pp. 393–94.

[6]*Ibid.*, p. 406.

[7]Letter of 20 March 1889 in Bunsen, pp. 228–29.

eral socialist, worker, and feminist congresses held in Paris. She was a constant and attentive listener at the June 14 Workers' Congress. Moreover, she was one of two Russian women delegates to the Congress of Women's Works and Institutions in Paris in July.[8]

Kovalevskaia continued corresponding with Maksim Kovalevskii, and there were visits back and forth between them. But at this time the relationship seems to have been one of friendship rather than anything deeper. She addressed him formally in her letters, and gave him as much information about her life and her thoughts as she gave other good male friends—Mittag-Leffler, her brother-in-law Aleksander Kovalevskii, Hansemann, Vollmar.

Kovalevskaia's apartment was the meeting place for many of those who came to see the Paris Exhibition that spring. Among her visitors was her cousin Mikhail, the "cousin Mishel" who had shared her algebra lessons at Palibino twenty-five years before. He was now a wealthy landowner; more intellectual than most, perhaps, but with no trace of the artistic aspirations of his youth. Kovalevskaia intended to write a novel based on his life—the theme was to be the transition from idealistic dreams to dull reality.[9]

Because Kovalevskaia was popular with Russian visitors as well as the permanent émigrés, she had some influence in the Russian colony. Consequently, when the satirical writer Saltykov-Shchedrin died in April 1889, Lavrov and the sociologist E. V. De Roberti asked her to use that influence to gather signatures for a telegram of sympathy to be sent to Saltykov's widow.

"Through thoughtlessness, which is a trait not only of the young, I eagerly accepted the task," Kovalevskaia wrote to Maksim, but "you cannot imagine what a mass of bourgeois hypocrisy I have seen in these last two days." To Sofia's sur-

[8]Petr Lavrov, *Russkaia razvitaia zhenshchina* (Geneva: Volnaia Russkaia Tipografiia, 1891), p. 1; E. G., "Zhenskii kongress," *Novosti i birzhevye vedomosti*, 16 July 1889; "Der erste weiblische Professor in Europa," *St. Jalles Stadtes Anzeiger*, 4 March 1891.

[9]A fragment of this appears as "Na vystavke," *VP 1974*, pp. 282–88.

prise, virtually none of her Russian friends would sign such an open statement of sympathy for the great progressive writer, in spite of their show of liberal and radical sentiments in private.

Excuses varied. One of Maksim's best friends, the positivist philosopher G. N. Vyrubov, refused to sign the telegram on the grounds that he was no longer a Russian but a Frenchman. Eventually, the Russian colony decided to form a committee to discuss whether they should make any public demonstration at all. This dithering further aroused Kovalevskaia's ire. "It will be twelve days since Shchedrin's death; maybe it's too early?" she wrote sarcastically to Maksim.[10]

Since the plan to send a group telegram fell through, Kovalevskaia decided to make her own memorial to Saltykov-Shchedrin. She wrote an article on the satirist's works, in which she tried to explain to European readers Saltykov's great importance for Russians. Drama was the same everywhere, she said, but satire was different for each country and each period. What made Saltykov so popular was his ability to satirize what most needed correction in the Russia of his day. The article was a fitting tribute to the great Russian writer.[11]

In spite of her disgust with certain members of the émigré community in Paris, Kovalevskaia found life in the French capital agreeable. There were always visitors from Russia or other émigré centers passing through, and the mathematical life there was far more intense than in Stockholm. Knowing that her five-year appointment in Sweden was ending, she began to search for a position in Paris or its suburbs, and even contemplated taking a second doctoral examination in France to facilitate this.

Weierstrass was alarmed at this new turn of events. He consulted with Mittag-Leffler and Hermite, but did not make his position known to Kovalevskaia until after the crisis had passed. Then, in mid-June, he summed up his views in an admonitory, fatherly letter. He said that although she had not asked for his advice, he would take that right upon himself as her oldest friend. Since there was little possibility of a French

[10]May 1889 letter in *VIP 1951*, pp. 304–05.

[11]"M. E. Saltykov (Shchedrin)," *VIP 1974*, pp. 221–30. The article was published in Swedish newspapers in June 1889.

university post for her, Sofia's decision to abandon Stockholm would mean a reduction in status. People would say that she herself had realized that women were unfit to be professors, and had taken a post in a *gymnasium* in recognition of that fact. "And after all, it was your idea once, and is hopefully still your idea now, to show through deeds that women have been alienated from the highest strivings of mankind because of prejudice," he reasoned.

Furthermore, Weierstrass felt that Kovalevskaia's plan to retake her doctorate in France would have far-reaching consequences. Even if she succeeded in getting herself admitted to the university, which he did not consider likely, there was no indication that she would be allowed to present her dissertation officially. And if she were allowed to take a French doctorate, then she would insult all those who had worked to make her Göttingen degree possible.

Finally, Weierstrass felt that Kovalevskaia was dreaming if she thought the French would give a Russian woman a chair of mathematics in Paris. She would be lucky, he thought, if they offered her a chair in the provinces. He advised her to abandon her plans for the present. Above all, she must do nothing that would give Stockholm the impression that she might not stay there.[12]

Weierstrass' calm, point-by-point discussion of the inadvisability of Kovalevskaia's plans for France gives no indication of the true extent of his anxiety about Sofia. Yet his letter was the outcome of several months of agitated correspondence with Hermite and Mittag-Leffler. The latter two also felt that Kovalevskaia would be unhappy in France for precisely the reasons Weierstrass had outlined in his letter. There was great prejudice against women in France, Hermite said. No Parisian faculty of mathematics would consider Kovalevskaia, and if a provincial university accepted her, she would be even more isolated mathematically than in Stockholm. The prominent mathematician Stieltjes, for example, had a position in Toulouse and, according to Hermite, was lonely and dissatisfied.[13]

Hermite was even more doubtful of Sofia's plan to teach in a

[12]12 June 1889 letter in *Briefe/Pis'ma*, pp. 147–48.

[13]Hermite, 26 January 1889 letter to Mittag-Leffler, IML.

French *gymnasium*. Her students would be vastly inferior to her Stockholm undergraduates, her teaching schedule would be demanding, prestige would be low. In short, he wrote to Mittag-Leffler, Sofia's idea of looking for a position in France was a bad one, and he and Bertrand were doing their best tactfully to discourage it.[14]

Mittag-Leffler, besides concurring with all of Weierstrass' and Hermite's objections to Kovalevskaia's plans, could add one more of his own: he did not want to lose Sofia as a colleague. He realized that Sofia was worried about the renewal of her appointment in June, and he knew that she was wise to inquire about other positions. Nevertheless, he was afraid that if news of her inquiries reached Stockholm, her already doubtful chances of a permanent professorship would be jeopardized further.[15]

The advice of Weierstrass, Mittag-Leffler, and Hermite was sound, and Kovalevskaia, who knew that they had her interests in view, seems to have followed it. At least, she did not make her dissatisfaction known to the university officials in Sweden. This was fortunate, because it was difficult enough for Mittag-Leffler to push through Sofia's professorship as it was.

Kovalevskaia's five-year appointment was due to expire May 11, 1889, and Stockholm University announced a concourse for the position. This in itself was an indication that Sofia would not have an easy time. The university administration evidently had decided that her case was not to be treated as a promotion, but rather as a fresh appointment decision.

Mittag-Leffler and Kovalevskaia were in suspense from the moment the concourse was publicized until it closed. The Prix Bordin and the excellent letters of recommendation sent by Hermite, the prominent Italian mathematician Eugenio Beltrami, and the Norwegian mathematician Carl Bjerknes had convinced all but one or two of the Stockholm professors of Kovalevskaia's scientific qualifications.[16] But Sofia and Gösta

[14]*Ibid.*, and Hermite, 4 and 18 February 1889 letters to Mittag-Leffler, IML.

[15]Mittag-Leffler, 28 December 1888 and 22 February 1889 letters to Weierstrass, IML.

[16]Mittag-Leffler, 15 April 1889 letter to Weierstrass; Hermite's, Beltrami's, and Bjerknes' letters of recommendation for Kovalevskaia, all in the IML.

knew that the appearance of a candidate of even moderate attainments would cause opponents of Kovalevskaia, Mittag-Leffler, and the Weierstrassian school to rally to his support.

Fortunately, other mathematicians apparently were reluctant to stake their qualifications against Kovalevskaia's, so there were no additional candidates of note. Moreover, Hermite and Beltrami both knew of the possibility of "attacks by malicious persons," as Beltrami phrased it. They endeavored to make their recommendations comprehensive and airtight.[17]

Even given the excellent planning of Mittag-Leffler, the absence of other candidates, and the support of Hermite, Beltrami and Bjerknes, Kovalevskaia's nomination encountered one more obstacle. Mittag-Leffler complained to Weierstrass:

> At the last moment I had to fight against the objection that one does not want a socialist appointed here. It was asserted that Frau Kovalevskaia is an intimate friend of the main socialist leader [Karl Hjalmar Branting] in Sweden, which is not entirely untrue. It was also asserted that she is the dominant influence on the socialists here, but this is entirely untrue.[18]

Whatever the truth of Kovalevskaia's involvement with Swedish socialists, Mittag-Leffler was able to soothe his colleagues. The vote eventually went overwhelmingly for Sofia. In mid-June 1889, Sofia Kovalevskaia was appointed to a lifetime professorship at Stockholm University. She was the first European woman in modern times to be so honored, and there would not be another after her for seventeen years. Sofia accepted the position, and placed in abeyance, for the time being at least, her plans to seek a post elsewhere.

To relax after the tensions of her reappointment battle, Kovalevskaia spent the summer of 1889 in France with her daughter Fufa, who had until then been staying with the Mittag-Lefflers in Sweden. Sofia rented a house in the Paris suburb of Sèvres, together with P. I. Iakobi and his family, and several

[17]Eugenio Beltrami, 16 May 1889 letter to Mittag-Leffler, IML; Hermite, 21 May 1889 letter to Mittag-Leffler, IML.

[18]Mittag-Leffler, 14 May 1889 letter to Weierstrass, IML.

other expatriate Russians. The group had many visitors. Iulia Lermontova came from Moscow, Lavrov and the famous doctor S. P. Botkin came from Paris, and Maksim Kovalevskii traveled from Nice.[19] The house rang with the sound of Russian voices raised in genial argument, and the time passed quickly and pleasantly.

Kovalevskaia's nephew, fifteen-year-old Iuri Jaclard, also visited Sèvres that summer. Sofia was worried about him, because his father Jaclard was thinking of remarrying. She requested that Victor allow her to raise Iuri with eleven-year-old Fufa. According to Sofia's daughter, Iuri would have gone to Stockholm with Sofia willingly. He loved her, and did not get along well with his father. Jaclard, however, refused his permission, so Iuri was allowed to visit but not live permanently with Sofia and Fufa.[20]

In mid-summer Kovalevskaia received word that she (along with Poincaré and others) had been honored by the French Ministry of Education by being named an "officer of public instruction." It was a purely honorary title, but she was pleased, because it indicated the esteem in which she was held by the French mathematical establishment. There was a certain amusement value in the document as well. The "M" had been obviously changed to "Mme," and an extra "e" was penned in at the end of "nommé." Clearly, Sofia was one of the first female officers.[21]

At the end of the summer Kovalevskaia and her daughter returned to Sweden, where Sofia resumed her lectures in a new capacity: lifetime professor. She had some difficulty settling into her duties again. She had been away for a semester and had not prepared her lectures in advance. But by this time she was experienced enough as a teacher to be able to improvise, so her classes went smoothly.

[19]S. Vl. Kovalevskaia, "Vospominaniia," *Pamiati*, p. 151.

[20]*Ibid.*, p. 152. Kovalevskaia's daughter and Iuri Jaclard corresponded for several years after her mother's death, but lost contact during the World War and Russian Revolution. Afterwards, when Sofia Vladimirovna tried to reestablish contact with her cousin, she was unable to find any trace of Iuri.

[21]13 July 1889 document from the French Ministry of Education, IML; Kovalevskaia, [22 July 1889 postmark on envelope] letter to Mittag-Leffler, IML.

During the fall and winter, Kovalevskaia worked on literary projects, sometimes with Anna Carlotta Leffler, sometimes alone. At that time, she finished her *Memories of Childhood*. In the presence of Anna Carlotta, Theresa Gyldén, Ellen Key, and others, she read them aloud, chapter by chapter, and they were translated from Russian as she read.[22]

The Swedish women felt that the reminiscences were certainly worth publishing. But out of fastidiousness they suggested that Sofia change them from the first person and disguise the names for the Swedish edition. This Kovalevskaia did, and in late 1889 the memoirs were published under the title *Ur ryska lifvet. Systrarna Rajevski* (From Russian Life. The Sisters Raevskii).[23]

Since Kovalevskaia's childhood memoirs appeared before her death only in two of the more inaccessible European languages (Russian and Swedish), it cannot be said that their publication catapulted her to literary fame throughout Europe. The reaction in Scandinavian intellectual circles, however, was favorable enough for Sofia to feel encouraged to pursue her writing. That winter she worked on several projects. These included a fictionalized biography of the radical Russian publicist Nikolai Chernyshevskii, the novella *A Nihilist Girl*, a novel with Maksim Kovalevskii as the hero, and a novel called *Vae Victis* (Woe to the Vanquished).[24]

Kovalevskaia and Anna Carlotta Leffler also attempted to edit a morose little play which Sofia had discovered among her late sister Aniuta's personal effects. Perhaps because of her gloomy mood at the time, Leffler had been charmed by the play

[22]The feminist and educator Ellen Key was at this time only at the beginning of her literary career. She taught at Fufa's school, and was the organizer of several feminist societies of which Kovalevskaia was a member. Others in their circle included the writer Amalia Vikstrom; a Miss Hedberg, who translated Sofia's memoirs into Swedish; and Iulia Lermontova, who visited from Moscow in the fall of 1889.

[23]The original Russian version was published in 1890 in the July and August issues of *Vestnik Evropy*.

[24]Of these, the last two were published in Kovalevskaia's 1893 collected works, edited by M. M. Kovalevskii. The Chernyshevskii manuscript had been known to exist, because both Leffler and Key mentioned it, but it was only discovered in 1953 by the Soviet writer L. A. Vorontsova. It is published in full in *VP 1974*, pp. 157–81.

when Sofia translated it for her. She felt that "the delineation of the characters was admirable, and throughout there lay in it a wonderfully deep, melancholy spirit." The two women worked enthusiastically to prepare the drama for the Swedish stage, but their efforts were in vain. Their friends felt that Aniuta's play was too depressing, too uniformly pessimistic to be successful in performance, and the project was abandoned.[25]

At the same time as Kovalevskaia was most involved in her literary endeavors, she was also attempting to arrange a position for herself in Russia. While in Paris during the previous spring, she had seen a distant relative of her mother, the governor of Saratov A. I. Kosich, to whom she confided her desire to return to her native country. He was impressed by her attainments, and touched by her obvious eagerness to work in Russia. Kosich wrote to the president of the Academy of Sciences, Grand Duke Konstantin Konstantinovich, asking him to help.[26]

Konstantin was not unsympathetic, or at least he had no wish to offend Kosich with an outright refusal. He turned the request over to the Academy Secretary, K. S. Veselovskii, who discussed it with several Academy members. In due time he sent Kosich a reply which was complimentary in tone, but negative in content. The Academy was proud that a country-woman had attained the high honors which had been given to Kovalevskaia, Veselovskii wrote:

> We are especially flattered [as fellow Russians] by the fact that Mme. Kovalevskaia has received a position as professor of mathematics at Stockholm University. The award of a university chair to a woman could occur only if everyone had an especially high and favorable opinion of her capabilities and knowledge. . . .
>
> [However], since access to teaching in our universities is completely closed to women, whatever their capabilities and knowledge, in our homeland there is no position for Mme. Kovalev-

---

[25]Leffler, "Sonya Kovalevsky," p. 275. The play was performed only once, in Denmark, and received uniformly unfavorable reviews because it was so depressing. There is a manuscript copy of the play, "V glushi" (In the Backward Provinces), in the IML. According to censors' notations on the front page, in 1883 the play was forbidden to be performed in St. Petersburg.

[26]Kovalevskaia, [autumn 1889] letter to A. I. Kosich, *VIP 1951*, pp. 305–06.

skaia as honorable and well-paid as that which she occupies in Stockholm.[27]

The refusal contained in this letter was clear enough. But so that Kovalevskaia would not have the slightest doubt of the decision, the mathematician and Academy member P. L. Chebyshev wrote to her personally as well. The letter was a masterpiece of shamefaced tact. The content, however, was the same as in the letter from Veselovskii to Kosich. There could be no university position for Kovalevskaia. "The only thing we can do is to be happy and proud that our countrywoman occupies a chair in a foreign university with such success," Chebyshev wrote.[28]

Chebyshev obviously felt guilty about his role in having to reject Kovalevskaia's plea for a Russian university position. After all, he was a close professional colleague and felt indebted to Sofia for publicizing his work in Western Europe.[29] Only days after he wrote the above letter he was instrumental in nominating his countrywoman for corresponding membership in the Imperial Academy of Sciences.[30]

The nomination was not immediately approved. Secretary Veselovskii pointed out that Kovalevskaia was a woman (no surprise!), and that her election was without precedent in the annals of the Academy. Did the academicians want to establish such a precedent merely to honor one woman? It seems that the academicians did, and on November 4, 1889 voted 20 to 6 in favor of the principle of admitting women as corresponding

[27]11 October 1889 letter from Veselovskii to Kosich, *VIP 1951*, p. 353. The academician most consulted in the matter was Chebyshev. He wrote what was essentially a rough draft, which was then amended by Veselovskii. (See "P. L. Chebyshev o Sof'e Kovalevskoi," *VIP 1951*, pp. 352–53 and p. 527, n. 352(3) to 353(4).)

[28]11 October 1889 letter from Chebyshev to Kovalevskaia, *ibid.*, pp. 351–52.

[29]There are several letters of thanks from him in the IML.

[30]The actual nomination reads: "In the physical-mathematical section of the Academy of Sciences/The undersigned have the honor to nominate for election as corresponding member in the mathematical section, doctor of mathematics, professor of Stockholm University, Sofia Vasilevna Kovalevskaia." The nomination is signed by P. Chebyshev, V. Imshenetskii, V. Buniakovskii. (*VIP 1951*, p. 354.)

members of the Academy. Three days later, the mathematical section approved Kovalevskaia's nomination with a vote of 14 to 3.[31]

Kovalevskaia was delighted with her election to the Russian Academy of Sciences, even though, as she acknowledged to her cousin Kosich, "corresponding member is no more than an honorary title and does not give me the possibility to return to Russia." What she really hoped for was that, now that the principled objection to electing a woman had been surmounted, the Academy "will no longer have an excuse not to choose me [for a full membership] just because I am a woman."[32] In Russia, as in many European countries, full membership in the Academy of Sciences carried a large salary, laboratory/office facilities, and more prestige even than a chair at a major university. As a full member of the Academy of Sciences, Kovalevskaia would have been financially able to remain in Russia.

Sofia spent the 1889–90 winter holidays partly with Anna Carlotta Leffler in Paris and partly with Maksim Kovalevskii at his villa in Nice. Anna Carlotta was amazed and alarmed by the variety of Kovalevskaia's acquaintances in Paris. Sofia knew scientists, mathematicians, writers, musicians, even bankers and the like. But she also knew "conspirators," in Leffler's delicate terminology: revolutionaries and socialists like Maria Jankowska-Mendelson, Lavrov, and S. A. Padlewski.[33]

At this time, George Kennan's firsthand observations of the brutality and horror of the Russian political exile system were appearing in British and American newspapers. Anna Carlotta described the effect of the news on Sofia and her friends:

---

[31]*VIP 1951*, pp. 513–14, n. 305(5). Besides the three signers of the nomination, Kovalevskaia's most vocal supporters were N. E. Zhukovskii and A. G. Stoletov. This was not the first instance of a woman being elected to an Academy of Sciences. The physicist Laura Bassi and possibly the mathematician Maria Gaetana Agnesi were members of the Bologna Academy in the mid-eighteenth century. (See Nikitenko, "Zhenshchiny-professora," *Russkaia mysl'*, No. 10 (1883), pp. 272 and 289; No. 11, p. 97.)

[32]Undated letter in *VIP 1951*, p. 306.

[33]Kovalevskaia wrote a paragraph-long biography of Padlewski on the flyleaf of one of her pocket notebooks at the IML.

"There was something deeply touching in the sorrow which the intelligence aroused in the Polo-Russian clique in Paris. It seemed as though its members had suffered personally. The sympathy of the clique with the Russian martyrs of the Tzar's cruelty was so strong that to all intents and purposes they were but one family."[34]

Despite Anna Carlotta's occasional misgivings about Sofia's friends, the two women enjoyed themselves in Paris. The pair entertained some of the most illustrious intellectual and scientific personalities then in the French capital, and Sofia was in top form. Sometimes, Sofia let herself be carried away by her high spirits. This type of mood was the cause of some coolness between her and her old acquaintance I. I. Mechnikov.

The physiologist Ilya Mechnikov had been introduced to Sofia by her late husband Vladimir in 1868. Sofia had not liked him much then, but she had reconciled herself to him for Vladimir's sake. She viewed Mechnikov with a certain amount of ironic amusement. He had courted her in Petersburg back in 1868–69, and had become miffed when she rejected him.[35]

Moreover, Sofia knew Mechnikov was not sympathetic to women's rights. When she was appointed to her five-year post in Stockholm, she had written to her brother-in-law: "I would be very interested to know what Mechnikov will say about this."[36] This was apparently a reference to Mechnikov's vociferously stated view that intellectual capabilities and genius were sex-related characteristics, as much man's exclusive property as beards.[37]

---

[34]Leffler, "Sonya Kovalevsky," pp. 278–80. George Kennan was an early American observer of the tsarist government who became sympathetic to the Russian revolutionary movement. His eyewitness accounts of Siberia were assembled into the two-volume *Siberia and the Exile System* (New York: Putnam, 1891). Kennan raised money in America for the Russian revolutionaries, and proudly claimed that he and the Russian political exile Sergei Stepniak-Kravchinskii were responsible for making American public opinion completely favorable to the revolutionaries. (Kennan, 15 October 1888 letter to Stepniak, TsGALI, f. 1158 (Sergei Mikhailovich Stepniak-Kravchinskii), op. 1, ed. khr. 299.) Kennan was a relative of the present specialist in Soviet affairs, also named George Kennan.

[35]"M. M. Kovalevskii o S. V. Kovalevskoi," p. 394.

[36]14 July 1884 letter in *VIP 1951*, p. 286.

[37]Semen Reznik, *Mechnikov* (Moscow: Molodaia Gvardiia, 1973), p. 98.

Mechnikov worked sometimes in the Pasteur Bacteriological Institute in Paris, and sometimes at the Russian Zoological Station outside of Nice. Although he and Sofia did not always get on well, he often visited her when she was in France, and was one of her circle of acquaintances that winter in Paris.[38]

Mechnikov had just made an important biological discovery—the role of white blood cells in the body's defense against infection—and was too smug and pleased with himself for Sofia's liking. The next time he visited, she blithely informed him that she and Anna Carlotta had been working on a play, the highlight of which was to be a ballet, entitled "the battle of the red and white blood cells." She described her fancy in meticulous detail. There would be dancers in red costumes battling dancers in white costumes; the whole thing was to be quite an extravaganza.

Instead of taking Sofia's idea as a joke, Mechnikov waxed indignant. The angrier he got, the more enthusiastically and absurdly Sofia described her ballet. Maksim Kovalevskii was astonished that Mechnikov took Kovalevskaia's raillery so seriously. Although he had known her only a few years, Maksim already realized that the simplest way to cut Sofia off in midflow was to laugh or agree with her.[39]

Mechnikov, on the other hand, had had twenty years of experience of Sofia's flights of fancy. Surely he should have known that his anger would only drive Kovalevskaia to more outrageous witticisms. But then, Mechnikov had never been particularly observant in anything outside of his own specialty.[40]

Toward the end of the Christmas holidays Sofia went off with Maksim Kovalevskii. Kovalevskaia was looking forward to her travels with Maksim. Almost immediately, however,

---

[38]*Pis'ma A. O. Kovalevskogo k I. I. Mechnikovu* (Moscow: AN SSSR, 1955), p. 145ff.

[39]"M. M. Kovalevskii o S. V. Kovalevskoi," pp. 394–95.

[40]The musician A. V. Goldenveizer, an intimate of the famous Leo Tolstoy, once described Mechnikov: "He talked a lot, and it was interesting and pleasant to be with him, but I did not sense any particular intrinsic worth in him." (Goldenveizer, *Vblizi Tolstogo* [Moscow: Izdatel'stvo Khudozhestvennoi Literatury, 1959], p. 267.) Mechnikov, incidentally, spent much of the visit to Tolstoy making derogatory comments about Kovalevskaia.

they began arguing, and the vacation Kovalevskaia had anticipated so happily was spoiled. With characteristic extravagance, she wrote to Anna Carlotta: "I see that he and I will never understand each other. I shall return to my work at Stockholm. In future my only consolation will be work."[41]

It is difficult to decide why Sofia's and Maksim's relationship became so stormy in 1890. Given the evidence available—Anna Carlotta Leffler's admittedly unreliable account, fragmentary diary entries, and Sofia's letters—it seems that Sofia's and Maksim's feelings for one another became more serious at the beginning of 1890, and they appeared to be heading toward marriage. They began to live together whenever Sofia's professional obligations would allow—at Christmas and Easter vacations and during the summer of 1890.[42] But naturally, such a haphazard arrangement caused tensions.

In May 1890, Kovalevskaia traveled to St. Petersburg. The academician V. Buniakovskii had died, and she had some hopes that she would be elected to fill his vacancy as a full member of the Imperial Academy of Sciences. Her reception in St. Petersburg was cordial in the extreme. She was officially welcomed back to the capital by the city duma (town council) and asked to give a speech. She was invited to sit in on the mathematics and physics examinations in the Higher Women's Courses, where the women gave her a photograph of the building signed by twenty-four students. She dined twice with the President of the Academy of Sciences Grand Duke Konstantin Konstantinovich and his wife, and visited members of the mathematical section P. L. Chebyshev and V. G. Imshenetskii.[43]

All in all, Kovalevskaia received a flattering amount of attention from her compatriots, and was encouraged to hope that a full Academy appointment was not far away. The Grand Duke Konstantin even went so far as to say "it would be nice" if Sofia

---

[41]Quoted in Leffler, "Sonya Kovalevsky," p. 284.

[42]Undated letter from Kovalevskaia to Leffler, probably early 1890, in *ibid.*, p. 284. On Kovalevskaia's previous visits to Maksim (summer 1888 in London and spring and summer 1889 in Nice and Sèvres), she seems not to have actually lived with him. That situation appears to have changed in 1890.

[43]Kovalevskaia, [20 May 1890 postmark] letter to Mittag-Leffler, IML.

could return to Russia as an academician. Chebyshev and Imshenetskii also were vaguely positive about the eventual possibility.

Nothing concrete was done at the time, however, because there was still opposition within the Academy both to the election of any female to full membership, and to the choice of Kovalevskaia in particular. The mathematician A. A. Markov had found what he claimed was a crucial mistake in two of her memoirs on the revolution of a solid body about a fixed point, and reactionary elements in the Academy were prepared to use his claim as an excuse to oppose her candidacy.[44]

Although Markov's "crucial mistake" turned out later to be an omitted case which was possible to fill in, at the time his criticisms were taken seriously. Under the influence of Markov's claims, several prominent mathematicians made it clear that they had doubts about Kovalevskaia.

Later, the Moscow Mathematical Society investigated Markov's allegations, and decided that they were "unsubstantiated and without foundation." The Society censured Markov, and declined to accept into the record of their meetings any more of his polemics against Kovalevskaia and other mathematicians. But this occurred too late to help Kovalevskaia. Markov was reprimanded in 1892, almost two years after her death.[45]

Because of Markov's complaints, Kovalevskaia's supporters had to move with caution. They decided not to pursue her nomination at this time. In fact, when Sofia asked Chebyshev whether she could attend the Academy's annual meeting as a corresponding member, he replied that "that was not a custom of the Academy."[46]

It is unclear whether Chebyshev discouraged Kovalevskaia's attendance because she was a woman, or because it was not customary for any corresponding member to be present. Probably, it was the latter. An added reason might have been the possibility that Kovalevskaia's nomination would be discussed

[44]Kovalevskaia, 22 December 1889 and [26 May 1890 postmark] letters to Mittag-Leffler, IML.

[45]Moscow Mathematical Society, 17 and 28 March and 17 November 1892 meetings, in *Matematicheskii sbornik*, 16 (1891–1892), pp. 834–45, *passim*.

[46]Kovalevskaia, 19 May 1890 diary entry, *VIP 1951*, p. 182.

informally by the Academy members during and after the meeting.

Kovalevskaia was disappointed that she was not to be elected to Buniakovskii's position immediately, but her reception at the hands of the Grand Duke Konstantin, Chebyshev and Imshenetskii was sufficiently cordial that she did not lose hope entirely. On balance, she was pleased with her visit, and intended to return in the spring of 1891 for an extended stay, when possibly Maksim Kovalevskii would be able to accompany her.[47]

After a short visit to Weierstrass in Berlin, Kovalevskaia spent the rest of the summer of 1890 traveling through Europe with Maksim. Her mood vacillated, and there were disagreements with Kovalevskii, as her letters and diaries attest. Most of the time, though, she seems to have been happy.

Kovalevskaia was in some sense traveling incognito. Because she was with Maksim, she did not want a fuss made over her in the capitals and resorts of Europe. She was something of a celebrity, and people seeing her name on a hotel register would have made her life unpleasant with their curiosity. Moreover, since everyone knew she was not married, she had no particular wish to give rise to more gossip than there was already.

Traveling incognito did not stop Kovalevskaia from seeing colleagues and friends, however. She transacted business for *Acta Mathematica*, saw her brother-in-law Aleksander in Amsterdam, and made a point of visiting her friend Hansemann, and Hermite and other mathematicians. Moreover, she worked on mathematics and literature. She wrote cheerful letters to her daughter and friends describing her scientific and/or literary plans.[48]

Kovalevskaia's *Memories of Childhood* were published in the July and August volumes of the Russian journal *Vestnik Evropy* (European Messenger). Although they lacked the polished style of established authors, they produced a sensation among the Russian educated reading public because of their content,

[47]Kovalevskaia, [December 1890] letter to Mittag-Leffler, IML.

[48]Kovalevskaia, [24 June and 10 July 1890 postmarks] letters to Mittag-Leffler, IML. See also diverse summer 1890 letters in the IML; and letters in *VIP 1951*, pp. 308–10 and Bunsen, p. 230.

implicit political orientation, and literary promise. The critic for *Severnyi vestnik* (Northern Messenger) wrote: "Our famous countrywoman should without doubt occupy one of the most prominent places among Russian authoresses . . . this work already contains all the indications of true literary power." Other reviewers were equally enthusiastic, and called for Kovalevskaia to continue the story of her life past her fifteenth year as soon as possible.[49]

Sofia was touched and pleased by the reception in Russia of her childhood memoirs. Her summer with Maksim and her friends in the mathematical community had refreshed her. She was eager to continue her scientific and literary projects, and returned to Stockholm in the fall of 1890 happy and full of energy. She wrote to A. N. Pypin:

> I am working much and with pleasure, because this summer I rested to my heart's content. . . .
>
> On my arrival here in Stockholm I was very pleasantly surprised to find letters from several Russian women whom I had never met. They told me how much they empathized with my memoirs and they insistently demand their continuation. These letters made me very happy and really convinced me to undertake a sequel: I at least want to describe my student years. Every minute I have free from mathematics I will now devote to that task . . .[50]

Kovalevskaia wrote to Maria Jankowska-Mendelson in much the same vein. She described several novels she was in the process of writing—*A Nihilist Girl*, the Chernyshevskii biography, and a novelette written in French which she had given to Jaclard for grammatical corrections.[51] She also discoursed at some length on Maria's involvement in a trial of Polish and

[49]B. B. Glinskii, 27 September 1890 letter to Kovalevskaia, IML; A. Volynskii [A. L. Flekser], "Literaturnye zametki," *Severnyi vestnik*, No. 10 (1890), p. 156. See also the favorable reviews in *Russkaia mysl'*, No. 8 (1890), pp. 390–93 and No. 9 (1890), pp. 438–44.

[50]Kovalevskaia, 29 September 1890 letter to A. N. Pypin, *VIP 1951*, p. 310. One of the editors of *Vestnik Evropy*, Pypin was a cousin of Chernyshevskii, and a friend of Kovalevskaia from the 1860s. He is mentioned frequently in her diaries.

[51]1 October 1890 letter in Mendelson, pp. 171–73.

Russian revolutionaries in Paris, and had obviously followed the details closely.[52]

While she worked on her own literary projects, Kovalevskaia was also actively engaged in promoting a taste for Swedish literature among Russians. She wrote several letters to the recording secretary of the *Northern Messenger*, B. B. Glinskii, advising him on which authors should be translated. Among them was her old enemy August Strindberg.[53] Strindberg had irritated her in the early days of her professorship in Stockholm by claiming that a female professor was a monstrosity, and that the Swedes had only invited her because of their famous gallantry toward the weaker sex.[54] In spite of his peculiar views on women, Sofia admired him as a writer, and recommended him to Glinskii.

In her letters to Glinskii, Kovalevskaia displayed knowledge of and interest in literary events in Russia as well. She inquired, for example, about whether Anton Chekhov's trip to Sakhalin had yielded any important fiction. (She knew Chekhov through Maksim Kovalevskii.) She also asked about the fate of several sentences in her article "Three Days in a Peasant University in Sweden," written for the November issue of *Northern Messenger*. She wondered whether the sentences had fallen afoul of the censor.[55]

With Kovalevskaia concentrating so much on literary matters, it is perhaps not surprising that ill-informed or ill-intentioned people decided that she was giving up mathematics. When people saw that she was talented in what might be considered a more "feminine" pursuit than mathematics, they assumed that she would forsake that "unwomanly" field for literature.

[52]*Idem.* The trial was that of a group of revolutionaries led by S. A. Padlewski, who had shot a retired Russian general living in Paris on the grounds that he was an agent of the tsarist police. Jankowska-Mendelson and her husband had to leave Paris after the trial because of the suspicion that they had aided Padlewski. (*VIP 1951*, p. 515.)

[53]24 October 1890 letter to Glinskii, *VIP 1951*, pp. 313–14.

[54]Kovalevskaia, 27 December 1884 letter to Mittag-Leffler, published in Leffler, "Sonya Kovalevsky," p. 219.

[55]Kovalevskaia, 20 November 1890 letter to Glinskii, *VIP 1951*, p. 315.

Kovalevskaia herself saw no contradiction between her two absorbing interests, however, and intended to pursue both. As she wrote to the belle lettrist A. S. Shabelskaia (Montvid) in the autumn of 1890:

> I understand that you are very surprised that I can work concurrently in literature and mathematics. Many who have never had the occasion to discover more about mathematics confuse it with arithmetic and consider it a dry and arid science. In reality, however, it is a science which demands the greatest imagination. One of the most prominent mathematicians of our century [Weierstrass] has said completely correctly that it is impossible to be a mathematician without having the soul of a poet . . .
>
> As for me, I have never been able to decide toward which I have the greater inclination: mathematics or literature. As soon as my mind becomes tired of purely abstract speculations, it immediately begins to turn toward observation of life, toward the urge to retell what I see. And conversely, at other times everything in life suddenly seems unimportant and uninteresting, and only the eternal, immutable laws of science attract me. It is very possible that I would have done more in either one of these fields if I had devoted myself exclusively to it, but at the same time I simply cannot abandon either one of them.[56]

Little time remained for Kovalevskaia to devote herself to either field, however.

In December 1890, Sofia left Stockholm to join Maksim Kovalevskii for the Christmas holidays at his villa in France. Her vacation was happy and relatively peaceful, and there seem to have been few arguments. Sofia enjoyed her trip so much that she would have asked for an extension of her holiday, as she wrote to Mittag-Leffler, were it not for the fear of jeopardizing her chances of obtaining a longer leave of absence to go to Russia with Maksim in the spring.[57]

The motive behind the longer leave of absence was two-fold. Mittag-Leffler, who was spending his holidays in St. Petersburg, wrote to Sofia that her candidacy for a full membership in

---

[56]Kovalevskaia, undated letter to Shabelskaia in "Nekrolog. Sofiia [sic] Vasilevna Kovalevskaia," *Kharkovskie gubernskie vedomosti*, 1 February 1891. Shabelskaia's 28 August 1890 fan letter to Kovalevskaia is in the IML.

[57]Kovalevskaia, [December 1890] letter to Mittag-Leffler, IML.

the Imperial Academy of Sciences was being discussed seriously. Mittag-Leffler had convinced Sofia that she needed to be in St. Petersburg when her nomination was proposed. Without her presence, he felt that there was no one "who would defend her interests with sufficient energy," as he wrote to Weierstrass.[58] Consequently, when Gösta informed Sofia of renewed interest in her in the Academy, she wanted to be in Petersburg to show how serious she was about desiring the Academy position.

The evidence also points to another reason for Sofia's wish for an extended leave in the spring. Apparently, Sofia and Maksim had decided to get married, and wished the ceremony to be performed in Russia, among their closest friends. If they married in the Russian capital, they could visit their Petersburg acquaintances, and Sofia could agitate for an Academy post at the same time.[59] Unfortunately, Kovalevskaia did not live until the spring.

Kovalevskaia left the south of France in late January 1891. Maksim accompanied her as far as Cannes, where she caught a cold. She needed to return to Stockholm in time for the spring semester, however, so she continued her journey in spite of her illness. She stopped in Paris and then Berlin to talk with Hermite, Picard, Weierstrass, Kronecker, and other colleagues. She apparently ignored her fast-worsening cold in her desire to crowd as much mathematics as she could into a few days.[60]

Kovalevskaia's colleagues in Paris and Berlin did not even notice her illness, so sparkling and full of plans was she that January. She discussed her hopes for an Academy of Sciences post, her Stockholm lectures, and a memoir by the mathematician Dirichlet. No one guessed that she was feeling ill—she charmed them all with her erudition and general liveliness.[61]

[58]*Ibid.*; Mittag-Leffler, 15 March 1890 letter to Weierstrass, IML.

[59]Kovalevskaia, [December 1890] letter to Mittag-Leffler and undated note from Anna Carlotta Leffler, introducing Maksim Kovalevskii as Sofia's "intended," IML; E. P. Kovalevskii [Maksim's nephew], "Cherty iz zhizni," in *M. M. Kovalevskii, 1851–1916*, p. 31; A. O. Kovalevskii, 4, 6, and 12 March 1891 letters to Lermontova, *VIP 1951*, pp. 357 and 518, n. 357(2).

[60]"M. M. Kovalevskii o S. V. Kovalevskoi," p. 392.

[61]February and March 1891 letters from Charles Hermite, Emile Picard, Hermann Hettner, and Karl Weierstrass to Mittag-Leffler, IML.

Since there was rumor of a smallpox epidemic in Copenhagen, Kovalevskaia, who was terrified of the disease, took an indirect, complicated route to Stockholm to avoid the afflicted city. This route entailed several mid-night changes of train, and since Sofia had no Danish money with her, she had to carry her bags herself. The weather was cold and rainy, making the tedious journey across the Danish islands even more exhausting and unpleasant than it was ordinarily.[62]

By the time Kovalevskaia returned to Stockholm, she was seriously ill, but she worked all day Thursday (February 5) and gave her scheduled lecture on Friday. Not wanting to neglect her fundraising obligations, she even went to a party at the Gyldéns' home that evening. But she was feeling so ill that she came home halfway through the event. The next morning she sent Mittag-Leffler a message scrawled on the back of one of her visiting cards asking him to send his doctor to her.[63]

Mittag-Leffler was alarmed, because Sofia was not the type to make much of small illnesses. He sent for a doctor immediately. The doctor recommended constant care, and a nursing sister was brought from the nearest hospital to be with Sofia night and day.

Kovalevskaia grew rapidly worse during that Saturday and Sunday, and was nursed by Ellen Key, Theresa Gyldén, and the nursing sister. Sofia seems to have had some idea that her illness was a mortal one, because she spent the last days of her life discussing the mathematical and literary works in which she was engrossed. It was almost as if she wished to record her planned accomplishments for posterity.

To Ellen Key and Theresa Gyldén, Kovalevskaia outlined her current and future writing projects. She said she had intended to complete her memoirs at least as far as her participation in the Paris Commune. She discussed a novel which was to be based partly on her father's life, and another, the above-mentioned biography of Chernyshevskii. This latter work was

---

[62]Leffler, "Sonya Kovalevsky," p. 289; A. O. Hauch, 10 February 1891 letter to Kovalevskaia, teasingly castigating her for not visiting him in Copenhagen on her way home from her Christmas vacation, IML.

[63]The visiting card is in the IML.

half complete, and Sofia dictated to Ellen Key the scene with which she had wanted to close the story.[64]

To Mittag-Leffler, Kovalevskaia gave the basic ideas of the mathematical project in which she was engaged. She had always been interested in applying Weierstrass' function theory to problems in mechanics and mathematical physics. Her works on light refraction and the rotation problem were two indications of her success in this direction. Now she outlined to her colleague further uses of elliptic and abelian integrals in mathematical physics. Mittag-Leffler was impressed. In fact, he felt that the work on which Kovalevskaia was embarked at the end of her life "would be the greatest she had yet written."[65]

Unfortunately, Kovalevskaia does not seem to have noted down her ideas, and Mittag-Leffler, believing her to be recovering, did not attend to the details of her presentation. The doctor had visited her that afternoon, and pronounced her out of danger. Theresa Gyldén and Ellen Key left to get some sleep, leaving the nursing sister in charge of the sickroom.

At her mother's insistence, Fufa was outfitted for a children's party. Sofia knew her daughter had been looking forward to it, and did not want to spoil Fufa's happiness. In fact, she derived satisfaction from overseeing Fufa's costume preparations. The party was on the following evening, however, so Fufa never did get to go.[66]

Sofia became much worse during the night, and after several painful hours of tossing and turning, lapsed into a coma. A telegram was sent off to Maksim Kovalevskii, and her friends were called, but Sofia did not regain consciousness. Early on the morning of Tuesday, February 10, 1891, Sofia Kovalevskaia died. She was just forty-one years of age.[67]

[64]Leffler, "Sonya Kovalevsky," pp. 290–91. Ellen Key's transcription of the last scene of the Chernyshevskii novel is in Leffler as well. No fragments of the Commune memoirs or the novel in which General Korvin-Krukovskii is featured have yet been found.

[65]*Idem.*

[66]*Idem*; Fufa, pencil draft of a letter to "Mama Iulia" (Lermontova), IML.

[67]The telegram to Maksim, along with letters from Lermontova, Mittag-Leffler, Fufa, and others to Maksim, were recently donated to the IML by P. Padievsky of Sannais, France. The cause of Kovalevskaia's death seems to have been pleurisy or pneumonia. She had a weak heart, but an autopsy after

News of Kovalevskaia's death stunned mathematicians and other intellectuals all over Europe. Charles Hermite found it almost impossible to believe that vital, active Sofia had expired. Nadezhda Stasova, a leading Russian feminist and former directress of the Petersburg Higher Women's Courses, wrote in her diary: "Kovalevskaia is dead! What a tragedy! She was not valued enough here." Anton Chekhov was told to "go to the devil" when he dared to make a pleasantry in a female acquaintance's presence in the days right after Sofia's death.[68]

Kovalevskaia's funeral was attended by many famous Swedish intellectuals, as well as large numbers of students and ordinary citizens. The church, the route of the funeral cortege, and the cemetery were thronged with people.[69] Telegrams and floral contributions poured in from all over Europe; two carriages were employed solely to transport bouquets and wreaths from the church to the cemetery.[70] Maksim Kovalevskii and Gösta Mittag-Leffler gave speeches at the gravesite, and the latter's brother Fritz Leffler recited a poem in honor of Sofia.[71]

There were memorial meetings for Kovalevskaia in Sweden, France, and in the mathematical societies of several countries, but by far the greatest tributes were paid to her by the intel-

---

her death determined that her heart condition did not cause her decease. (Leffler, "Sonya Kovalevsky," p. 292; "Pokhorony S. V. Kovalevskoi," *Novoe vremia*, 13 February 1891; "M. M. Kovalevskii o S. V. Kovalevskoi," p. 392.)

[68]Hermite, 14 February 1891 letter to Mittag-Leffler, IML; Stasov, *Nadezhda Vasilevna Stasova*, p. 395; A. P. Chekhov, 31 January 1891 letter to A. S. Suvorin, *Pis'ma A. P. Chekhova* (Moscow: Knigoizdatel'stvo pisatelei, 1915), vol. III, p. 182.

[69]Although Kovalevskaia was not at all religious, and was in fact antireligious, her funeral service was Russian orthodox. This seems to have been the doing of her Swedish friends.

[70]"Pokhorony," *Novoe vremia*, 13 February 1891; *Dagens Nyheter*, 17 February 1891; and a list of telegrams and floral offerings in the IML. Telegrams and flowers came from the mathematicians Hermite, Hensel, Kronecker, Weierstrass, and others, the Swedish and Russian Academies of Sciences, Saratov governor A. I. Kosich, and universities, students' groups, and women's organizations all over the Russian empire.

[71]The speeches of Kovalevskii and Mittag-Leffler are in "Pokhorony," and Leffler's poem is in *VIP 1951*, p. 317. Lavrov gave a speech at a 6 April 1891 memorial meeting in Paris for Sofia. It was published as *Russkaia razvitaia zhenshchina. V pamiati S. V. Kovalevskoi*. Other articles of an obituary or commemorative nature are listed in the bibliography.

ligentsia, especially the women, of her own homeland. It was they who erected the monument over her grave in Stockholm, they who collected money for a mathematical scholarship in her name, they who most cherished her memory as a mathematician, writer, and champion of women's rights.[72]

Years after her death, Kovalevskaia's grave was still a place of pilgrimage, where tribute could be paid to a great woman. A Russian visitor to Stockholm in 1900 noted three floral offerings adorning her tomb: one from the Fundraising Committee of the Higher Women's Courses, one from the Russian Women's

[72]There were memorial meetings on the fifth, tenth, and twenty-fifth anniversaries of Kovalevskaia's death. See newspaper accounts for 1891, 1896, 1901 and 1916 meetings listed in the bibliography.

Kovalevskaia's monument in Stockholm was financed through contributions collected by the Fundraising Committee of the Petersburg Higher Women's Courses. The former Directress of the Courses, N. V. Stasova, initiated the drive for a monument and scholarship in Sofia's name, and wanted to do something for Fufa as well. Kovalevskaia's daughter eventually did receive a stipend from the Higher Women's Courses.

Stasova and Iulia Lermontova had a slight tussle with Mittag-Leffler over the type of monument to be put on Kovalevskaia's grave. Mittag-Leffler held out for a grandiose obelisk with Sofia's bust on top. The Russian women wanted a simple black marble cross. In the end, Stasova and Iulia won out, because Iulia insisted that Kovalevskaia would never have wanted her own bust on her grave. (Lermontova, 18 March 1892 letter to Stasova, TsGALI, f. 473, op. 1, ed. khr. 8, d. 25.)

After Kovalevskaia's death six parties offered to raise Fufa: Sofia's cousin A. I. Kosich, governor of Saratov; Fufa's godfather, the physiologist Ivan Sechenov and his wife Maria Bokova-Sechenova, who had been friends of Sofia since the 1860s; Iulia Lermontova; Sofia's brother-in-law Aleksander Kovalevskii; the astronomer Hugo Gyldén; and Maksim Kovalevskii, who even offered to adopt Fufa and make her his heir. (See letter, in *VIP 1951*, pp. 356–57; and documentation, letters, and rough drafts in the Kovalevskaia papers of the IML.) There was quite a prolonged dispute, with Mittag-Leffler ranged on the side of Maksim's claim and the physicist Hansemann defending Lermontova. Lermontova's claim was considered the strongest, though, so after Fufa completed her primary school in Sweden (she lived with the Gyldéns), she returned to Moscow. She went to the Higher Women's Courses in Moscow, became a doctor, and worked in the Red Cross in the Soviet Union and abroad. After her retirement, she worked as a medical librarian and translator from Swedish. She died childless in the Soviet Union at the age of seventy-four. (See S. Vl.'s letters to Mittag-Leffler and Maksim Kovalevskii in the IML (she corresponded with them both until their deaths); [Polubarinova] Kochina, *Sofia Vasilevna Kovalevskaia*, pp. 281–82; Amalia Fahlstedt, "Sonja Kovalevskys minne," *Idun*, 26 September 1896. Fahlstedt's article has a picture of the eighteen-year-old Fufa—she looked very like her mother.)

Mutual Aid Society, and one pot with a few wildflowers and no inscription—a touching indication of some nameless person's homage.[73] The bouquets were added evidence of what Maksim Kovalevskii had said at Kovalevskaia's funeral: that she always was and always would be "a faithful and devoted ally of young Russia; a Russia peaceful, just, and free, the Russia to which belongs the future."[74]

[73]Iv. Zabrezhnev, "U finnov i shvedov," *Novoe vremia*, 15 September 1900. It is worth noting that even in 1900 Zabrezhnev felt that his visit to Kovalevskaia's grave was a natural part of his trip to Scandinavia and that it needed neither explanation nor justification.

[74]M. M. Kovalevskii, "Rech' na mogile," *VIP 1951*, p. 407. The inscription on Kovalevskaia's monument reads (in Russian): "To Professor of mathematics Sofia Vasilevna Kovalevskaia from her Russian friends and admirers." The dedication of this monument in 1896 was an emotional event. (See Lermontova's letters to Stasova (who had planned to attend the dedication in spite of her age and ill health, but died several months before), TsGALI, f.473, op. 1, ed. khr. 8, d. 25; "Pamiatnik S. V. Kovalevskoi," *Novoe vremia*, 11 October 1896; Fahlstedt, "Sonja Kovalevskys minne," *Idun*, 26 September 1896.)

thirteen

KOVALEVSKAIA'S MATHEMATICS

Sofia Kovalevskaia published only ten mathematical papers during the course of her short lifetime, and of those, two are French and Swedish editions of the same work.[1] The papers were written during two periods: from 1871 to 1874, when she was studying with Karl Weierstrass in Berlin and was interested in theoretical problems in analysis; and from 1881 to the end of her life ten years later, when she lived in various parts of Europe and was interested in mechanics and mathematical physics.

The two periods of research at first seem separate, not only because of the seven-year hiatus in scientific productivity, but also because Kovalevskaia's research switched from pure to more applied areas. As Kovalevskaia's biographer and fellow mathematician Elizaveta Litvinova points out, however, "there is an internal connection between the works of the first and second periods."[2] That connection is supplied by Kovalevskaia's constant use of function theoretic techniques developed by Weierstrass and employed by her for the resolution of practical as well as theoretical problems.

Mathematicians and historians of science would agree that Kovalevskaia's most significant contributions were two: her proof of the theorem in partial differential equations now re-

---

[1] The works have been collected, translated into Russian, and published with editorial commentary by P. Ia. Polubarinova-Kochina in S. V. Kovalevskaia, *Nauchnye raboty* (Moscow: AN SSSR, 1948).

[2] Litvinova, p. 87.

ferred to as the Cauchy-Kovalevskaia Theorem, and her work on the revolution of a solid body about a fixed point. The former was one of the three papers she presented to Göttingen University as her doctoral dissertation in 1874. The latter was the research for which she won the Prix Bordin of the French Academy of Sciences in 1888.

Kovalevskaia's first important work was in the area of partial differential equations. These equations describe situations which occur frequently in the physical world, any time one has a function of several variables. Atmospheric temperature, for example, can be considered as a function depending on four variables: latitude, longitude, altitude, and time. A partial derivative of such a function is the rate of change of the function with respect to one of the variables. For example, the partial derivative of temperature with respect to time is the rate of change of temperature if only time varies and the other variables (i.e., the location) remain constant.

Since multivariable functions occur constantly in applications, so do partial differential equations—equations involving partial derivatives—because often the most easily obtainable information about a function involves relations between its rates of change with respect to certain of its variables. It is for this reason that partial differential equations are considered a basic field of pure and applied mathematics. Almost all major university mathematics departments contain at least one expert in the subject.

In "Zur Theorie der partiellen Differentialgleichungen" (Toward a Theory of Partial Differential Equations) Kovalevskaia proved that under certain conditions, there exists one and only one solution to a given partial differential equation.[3] The French mathematician Augustin Cauchy had posed the problem in 1842 and had given a solution, but neither Weierstrass nor Kovalevskaia was aware of his work in 1873–74.

In fact, mathematicians did not become generally aware of Cauchy's work on the question until 1875, when the Frenchman Gaston Darboux published a paper more or less duplicating Kovalevskaia's results. In the ensuing priority dispute,

---

[3]First appeared in *Journal für die reine und angewandte Mathematik* [Crelle's Journal], 80 (1875), pp. 1–32.

waged by Weierstrass on Kovalevskaia's behalf and by Hermite on Darboux's behalf, Cauchy's solution was uncovered.[4]

Kovalevskaia's results and the simplicity of her exposition were greatly admired by specialists at the time. The French mathematician Henri Poincaré, for example, was impressed by the elegance of Sofia's paper. In his famous memoir on the three body problem, he wrote: "Madame Kovalevsky significantly simplified Cauchy's method of proof, and gave the theorem its final form."[5]

Karl Weierstrass was pleased not only with his pupil's final results, but also with some side avenues she explored in her research. Kovalevskaia had examined equations for heat conduction, and discovered that certain partial differential equations have no analytic solutions even when they have "formal power series" solutions. On May 6, 1874, Weierstrass wrote:

> You see, dearest Sonia, that your observation (which seemed so simple to you) about the peculiarity of partial differential equations—that an infinite series may formally satisfy such an equation without converging for any system of values of its variables—was for me the starting point for interesting and illuminating research. I hope my student will continue to express her gratitude to her teacher and friend in this same way [in future].[6]

The Cauchy-Kovalevskaia Theorem is basic to the theory of partial differential equations. Charles Hermite called Kovalevskaia's formulation "the last word" on the subject, and said that the paper would be the point of departure for all future research in partial differential equations.[7] The modern Soviet

[4]See 21 April 1875 letter from Weierstrass to Kovalevskaia in *Briefe/Pis'ma*, pp. 67–8.

[5]Henri Poincaré, "Sur le problème des trois corps et les équations de la dynamique," *Acta Mathematica*, 13 (1890), p. 26. (This was the paper which had won for Poincaré the Oscar Prize of the Swedish Academy of Sciences.)

[6]*Briefe/Pis'ma*, p. 38. See also Kovalevskaia's 15 July 1885 letter to Mittag-Leffler, IML. Here, she notes that neither Cauchy nor Darboux had remarked on what she considered the most important point, which related to the conditions under which certain partial differential equations are integrable.

[7]Hermite, 22 March 1886 and 21 May 1889 letters of recommendation for Kovalevskaia, IML.

mathematician O. A. Oleinik agrees, and says: "Kovalevskaia's work marked the beginning of the development of a general theory of partial differential equations."[8]

The other theme of Kovalevskaia's most important research concerned the classical problem of the revolution of a solid body about a fixed point.[9] Examples of motion of this type are pendula, tops and gyroscopes. Sofia had been interested in this problem almost from the start of her mathematical career, and had always felt that it would be possible to resolve it with the help of abelian functions. (These are functions defined using certain types of definite integrals; in some sense they are generalizations of the familiar trigonometric functions.)

Mathematicians had been working for over one hundred years on the task of analyzing the motion of a solid body about a fixed point. Two classical cases had been studied by Euler, Lagrange, Poisson and Jacobi. Kovalevskaia analyzed the third classical case of this problem. The type of solid body she studied was the most complicated and the most difficult to analyze.

What amazed Kovalevskaia's contemporaries most about her solutions was their simplicity and elegance. Her reasoning was so clear, her grasp of abelian functions so complete, that the steps of her argument flow one from the other with ease. The Russian specialist in mechanics N. E. Zhukovskii remarked: "The analysis she uses is so simple that, in my opinion, it would be worthwhile to include it in analytic mechanics courses."[10] The mathematician P. A. Nekrasov agreed, and

---

[8]O. A. Oleinik, "Teorema S. V. Kovalevskoi i ee rol' v sovremennoi teorii uravnenii s chastnymi proizvodnymi," *Matematika v shkole*, No. 5 (1975), p. 5.

[9]Kovalevskaia's three papers on this subject are: "Sur la problème de la rotation d'un corps solide autour d'un point fixe," *Acta Mathematica*, 12 (1888–1889), pp. 177–232; "Sur un propriété du système d'équations differentielles qui definit la rotation d'un corps solide autour d'un point fixe," *Acta Mathematica*, 14 (1890), pp. 81–93; "Mémoire sur un cas particulier du problème de la rotation d'un corps pesant autour d'un point fixe, où l'intégration s'effectue à l'aide de fonctions ultraelliptiques du temps," *Mémoires présentés pars divers savants à l'académie des sciences de l'Institut de France*, 31 (1894), pp. 1–62.

[10]N. E. Zhukovskii, "O trudakh S. V. Kovalevskoi po prikladnoi matematike," *Matematicheskii sbornik*, 16 (1891), p. 18. Her method is, in fact, used in some mechanics texts even today.

said: "Her clever timing is reflected in the adroitly devised gradual transition from the simple to the more complex, in her masterly ability to bring the very difficult [closer] to the less difficult."[11]

Modern mathematicians, on the other hand, are impressed not so much by the simplicity of Kovalevskaia's solution to the rotation problem, as by the general applicability of one of the techniques she used in her rotation papers. Kovalevskaia developed a method to find necessary conditions for algebraic integrability. Her "asymptotic method," it turns out, is not restricted to the motion of rigid bodies only, and is being used by mathematical physicists for other problems today.[12] This will be discussed further below.

Kovalevskaia's other works are of less importance than the ones discussed above, but nevertheless they deserve mention. One of her three dissertation papers was on the reduction of abelian integrals to simpler elliptic integrals.[13] For this she employed some of Weierstrass' newest results and, according to Litvinova, "solved a very difficult problem with great skill and proficiency."[14]

Kovalevskaia's third dissertation, on the shape of Saturn's rings, was on a problem in classical astronomy.[15] Operating on the theoretical assumption that the rings are fluid, she improved upon Laplace's previous model of the shape of the rings. Laplace had believed that the rings are ellipsoidal; Kovalevskaia decided that they must be egg-shaped ovoid and

[11]P. A. Nekrasov, "O trudakh S. V. Kovalevskoi po chistoi matematike," *Matematicheskii sbornik*, 16 (1891), p. 35. See Kovalevskaia's 1888 letters to Mittag-Leffler, explaining her work on the rotation problem, IML.

[12]Dr. Mutiara Buys, 15 June 1983 letter to the author.

[13]"Über die Reduction einer bestimmten Klasse von Abel'scher Integrale 3-en Ranges auf die elliptische Integrale," *Acta Mathematica*, 4 (1884), pp. 393–414.

[14]Litvinova, p. 88. This was the paper Kovalevskaia presented at the 1879 Congress of Natural Scientists in St. Petersburg.

[15]"Zusätse und Bemerkungen zu Laplace's Untersuchung über die Gestalt des Saturnringes," *Astronomische Nachrichten*, 111 (1885), pp. 37–48.

oriented in a certain way.[16] Since the rings in reality are known to be composed of solid rather than liquid matter, Sofia's actual conclusions were not of lasting significance. However, the power series method she used was later applied to other problems by Poincaré, so the paper did have some use.[17] Moreover, as usual, her method of exposition was so sure and confident that some mathematicians read the work for the sheer intellectual pleasure of seeing how an expert dealt with computational difficulties.[18]

On the subject of refraction of light in a crystalline medium, Kovalevskaia wrote three papers, which were of interest to her contemporaries.[19] In these, she pointed out several mistakes which the physicist Lamé had made in his presentation of the problem. Unfortunately, she overlooked his most important error, which invalidated both Lamé's and her own work.[20] However, as Polubarinova-Kochina points out, the first of Kovalevskaia's articles on this subject was "still a valuable paper, because it contained an exposition of Weierstrass' theory for integrating partial differential equations—Weierstrass himself often delayed with the publication of his research."[21]

Kovalevskaia's last published mathematical work was a short

---

[16]Roger Cooke, manuscript of *The Mathematics of S. V. Kovalevskaya*. As it happens, a recent space probe has determined that Saturn's rings do have approximately the same shape as Kovalevskaia's model predicted. But since she was assuming liquid conditions, and the rings are made of solid matter, this seems to be no more than a curious coincidence.

[17]Polubarinova-Kochina, "On the Scientific Work of Sofya Kovalevskaya," p. 247.

[18]Eugenio Beltrami, 18 May 1889 letter of recommendation for Kovalevskaia, IML.

[19]"Über die Brechung des Lichtes in crystallinischen Mitteln," *Acta Mathematica*, 6 (1885), pp. 249–304; "Sur la propagation de la lumière dans un milieu cristallisé," *Comptes rendus de l'Académie des Sciences*, 98 (1884), pp. 356–57; "Om ljusets fortplanting uti ett kristallinskt medium," *Of versigt af Kongl. Vetenskaps-Akademien Forhandlinger*, 41 (1884), pp. 119–21. (The latter two are summaries of the results of the first.)

[20]Vito Volterra announced his discovery of this error in his 3 June 1891 letter to Mittag-Leffler, IML.

[21]Polubarinova-Kochina, "On the Scientific Work of Sofya Kovalevskaya," in *Russian Childhood*, pp. 247–48.

article of a theoretical character.[22] In it, she gave a simplified proof of a theorem of Bruns in potential theory, using her own dissertational research in partial differential equations (the Cauchy-Kovalevskaia Theorem).[23]

As significant as Kovalevskaia's published works were, they do not tell the whole story of her impact on mathematics. It is possible that Sofia's main importance lies in what she did in a general mathematical sense even more than in what she wrote.

Kovalevskaia had been trained by Weierstrass in his school of purely theoretical analysis and function theory, and had acquired a complete grasp of the most abstruse of his methods. She had also been accepted as a colleague by most of the great mathematicians of Western Europe—an acceptance achieved in spite of Sofia's sex.

It is worth noting that Kovalevskaia only once mentions being discriminated against by mathematicians because she was a woman: when the mathematician Kummer voted against her admission to Berlin University in 1870. Other than that one incident, skirmishes with prejudice seem to have been brought on by university officials, other kinds of professors (the chemist Bunsen and the botanist Wittrock, for example), the tsarist government, and the public at large. By contrast, mathematicians on the whole were tolerant and supportive.

This attitude on the part of mathematicians continued. In 1897, one hundred German professors were sent a questionnaire, asking if they would vote to admit women to the university on an official basis. The most favorable to the proposal were mathematicians, the least favorable were historians. The mathematicians' positive response was attributed to the respect in which the work of Kovalevskaia and Sophie Germain (who in 1816 had won the Grand Prix of the French Academy of Sciences for her work on the elasticity of metals) was held by specialists.[24]

Russian mathematicians also had relatively little objection to

[22]"Sur un théorème de M. Bruns," *Acta Mathematica*, 15 (1891), pp. 45–52.

[23]Polubarinova-Kochina, "Nauchnye raboty S. V. Kovalevskoi," *Uspekhi matematicheskikh nauk*, 5, No. 4(38) (1950), pp. 13–4.

[24]*Le Journal de Marseilles*, 8 April 1897.

Sofia on the basis of her sex. As was mentioned above, their mistrust stemmed more from hostility to the purely theoretical, abstract tendency of the Weierstrassian school, and from their distrust of German-educated Russians. But, partly because Kovalevskaia had had her early mathematical training in Russia, and partly because she was such a good expositor and popularizer, she was eventually able to overcome the skepticism of her countrymen.

In spite of her Weierstrassian tendencies, Russian mathematicians discovered that Sofia had never lost her respect and intuition for practical problems in mechanics and mathematical physics. As a result, she had much in common with the specialists in her own country, who liked a down-to-earth approach, even in non-applied areas of mathematics such as number theory. Kovalevskaia was therefore able to serve as an intermediary between the mathematicians of Western Europe and those of Russia.

In the years before Kovalevskaia's 1879 talk at the Congress of Natural Scientists in St. Petersburg, Russian mathematicians had related to the new developments of Weierstrass and his school with suspicion. They considered the new brand of analysis too abstract and ungrounded in practical reality to be appealing. They either refused to read or deliberately misunderstood research in that area.

From the time of her return to Russia in 1874 until her death in 1891, Kovalevskaia served as a liaison between the school of Weierstrass and the inward-looking but capable mathematicians of the Russian empire. At first, P. L. Chebyshev, P. A. Nekrasov, N. E. Zhukovskii and others were reluctant to listen to Sofia's ideas. They were prejudiced against "the German tendency," partly for nationalistic reasons. In many situations, they favored French mathematics and French mathematicians over their German counterparts. Sofia kept insisting on the usefulness of Weierstrass' function theory, however, and by the end of 1879, Chebyshev at least was receptive enough to the new ideas to encourage her to talk about them at the Russian Congress of Natural Scientists.

Kovalevskaia chose to present her second dissertation paper. In this work, Sofia used Weierstrass' techniques to express a certain class of abelian integrals in terms of less complicated elliptic integrals. Fortunately for the immediate future of Weierstrass' ideas in Russia, Kovalevskaia was an excellent popularizer.[25] Her exposition was, as usual, so clear and elegant that she made intricate computations seem natural and beautiful.

As a result of Kovalevskaia's talk, Elizaveta Litvinova says that Russian mathematicians "began to relate [to Weierstrass' method] with less hostility and more interest." They asked Kovalevskaia numerous questions, and sought her assistance in guiding them to relevant mathematical literature published in the West. Weierstrassian analysis had finally made its way into Russia, and Sofia was able to note the change with pride. "To my great surprise as well as satisfaction . . . he [Chebyshev] spoke with respect of the Berlin school," she wrote to Mittag-Leffler in 1881.[26]

After Kovalevskaia moved to Stockholm in 1883 and became an editor of *Acta Mathematica*, her Russian colleagues viewed her more than ever as their personal representative in the West. She was besieged by letters with requests and questions from all corners of the Russian empire. Professors wrote about their own work and that of their students, students requested help on their dissertations, beginners asked what they should read.[27]

Even such an eminent mathematician as P. L. Chebyshev, who had been publishing in French for forty years, asked Kovalevskaia for help in sorting out the advances of Western mathematics.[28] Chebyshev was a notoriously poor corre-

[25]Her colleagues often turned to her for clear explanations of Weierstrass' ideas. See letters from Poincaré, Runge, Selivanov, and others in the IML. Weierstrass himself often asked Sofia to help clarify his own ideas. (See *Briefe/ Pis'ma, passim.*)

[26]Litvinova, p. 88. See also V. E. Prudnikov, *Pafnutii Lvovich Chebyshev 1821–1894* (Leningrad: Nauka, 1976), p. 227ff; Kovalevskaia, undated [1881] letter to Mittag-Leffler, IML.

[27]There are several folders of such letters in the IML.

[28]There are several unpublished letters from Chebyshev to Kovalevskaia in the IML.

spondent. Hermite once wrote to him in exasperation: "I realize that it is your privilege *not* to answer letters . . ."[29]

Yet Chebyshev managed to write to Kovalevskaia on a regular basis. In fact, about one-third of the (preserved) mathematical letters he wrote during a long and illustrious career were addressed to her. When one considers that their correspondence was only during the last ten years of Kovalevskaia's life, the importance Chebyshev attached to his contact with her becomes even more apparent.[30]

Kovalevskaia's service as a conduit of the newest theoretical ideas and approaches is less tangible than her own papers, but its importance should not be underestimated. Although many Russian mathematicians of her time were skillful, inventive specialists, they worked almost in ignorance of the developments in the rest of Europe. Kovalevskaia was instrumental in changing that, and it may well be that this should be recognized as one of her greatest, most enduring contributions to the history of mathematics.

In evaluating Kovalevskaia as a mathematician, one should take into account what her contemporaries said about her, what influence she had on the direction of mathematical research in her own time and afterwards, and what impact, if any, her ideas are having on the course of mathematics today. It should be remembered that she died soon after her forty-first birthday. Like her former teacher Weierstrass, however, she was one of those mathematicians who become more productive as they age rather than less so. Her colleagues were unanimous in saying that she died in the full flower of her creative abilities—such phrases were used repeatedly in the letters sent to Mittag-Leffler after Sofia's death.[31] But since there is no use speculating on what Kovalevskaia would have accomplished

[29]Quoted in Chebyshev's *Polnoe sobranie sochinenii* (Moscow: AN SSSR, 1951), vol. V, p. 412.

[30]See Chebyshev's collected letters in *ibid.*, vol. V, p. 412ff. Several letters in the IML are not included in the volume.

[31]See letters from Hermite (26 February 1891), Beltrami (20 March 1891), Picard (15 February 1891) and others to Mittag-Leffler in the IML; and the *Matematicheskii sbornik* commemorative issue cited above.

had she lived longer, one can only comment on her actual achievements.

Kovalevskaia's contemporaries praised her work extensively, although no one would have called her a pathbreaker. Henri Poincaré found her exposition elegant, her methods polished, her analysis clear. He divided all mathematicians into two categories, and characterized her as a "logical" or "analytical" as opposed to an "intuitive" or "geometric" mathematician. By this, Poincaré meant that she and other "analytical" mathematicians (who included Weierstrass and to a lesser extent Hermite) saw the structure of their research without constructing pictures in their minds. They pursued a certain line of reasoning to its logical conclusion without needing visually to imagine steps along the way. On the other hand, "geometric" mathematicians (who included Riemann, Bertrand, and Lie) could only reason with the help of their imagination, using pictures they set up in their minds. Kovalevskaia, as an analytical mathematician, was one of those who

> perceive at a glance the general plan of a logical edifice, and that too without the senses appearing to intervene. In rejecting the aid of imagination, which . . . is not always infallible, they can advance without fear of deceiving themselves. Happy, therefore, are those who can do this without aid! We must admire them, but how rare they are!

But Poincaré added that among the analytical mathematicians there will be few pathbreakers, because "sensible intuition is in mathematics the most useful instrument of invention."[32]

Other mathematicians agreed with this general evaluation. Elizaveta Litvinova wrote:

> Poincaré and other first-class mathematicians followed the results of her work with great interest. Nevertheless, it is generally believed that Kovalevskaia was not one of the geniuses of mathematics—she did not institute any revolution, but she was without question equal to the most talented male mathemati-

---

[32]Henri Poincaré, *The Foundations of Science* (1913) (Lancaster, Pa.: The Science Press, 1946), pp. 212, 221. Kovalevskaia probably would have disagreed with this description; she thought she was intuitive and geometric.

cians of our time. She probed deeply into existing methods, used them in the cleverest ways, shared them with others, and developed them. She made completely new, brilliant discoveries and easily handled the greatest difficulties.[33]

The Moscow mathematician P. A. Nekrasov used almost the same words in his commemorative speech: "I don't want to exaggerate the extent of her mental gifts . . . but I must say that S. V. Kovalevskaia was without question equal to the most talented male mathematicians of our time . . ."[34]

It should be noted that Nekrasov and Litvinova both meant "our time" to signify a "mathematical generation," i.e., about five to ten years. At the time of her death, Kovalevskaia was indeed considered the equal of anyone in her generation. This included even Poincaré, Picard, and Mittag-Leffler. It should also be pointed out that Nekrasov later apologized in the pages of *Matematicheskii sbornik* for being so grudging in his praise of Kovalevskaia. He was under the influence of Markov's attacks on her work at the time, he said.[35]

Not everyone, however, was willing to acknowledge Kovalevskaia's importance as a mathematician. Since the quality of her work could not be denied, some people decided that it had all been done or closely supervised by Weierstrass. One of the main propagators of such ideas was Sofia's old acquaintance and spurned suitor, the physiologist Ilya Mechnikov.

Mechnikov was not a mathematician, but as an eminent scientist he circulated among mathematicians and physicists in Europe and Russia. His information was generally believed, because he was a close friend of the Kovalevskii family, and so was thought to speak with authority. He constantly told the following story: "She worked under the direction of the famous Berlin mathematician Weierstrass, who was already middle-aged at the time. He became infatuated with her and under the influence of his infatuation he gave her the idea of her work, which she merely carried out."[36]

[33]Litvinova, p. 90. Litvinova goes on to say that had Kovalevskaia lived longer, she might have been a pathbreaker.

[34]Nekrasov, "O trudakh," p. 36.

[35]*Matematicheskii sbornik*, 16 (1891–1892), pp. 844–45.

[36]Quoted in Goldenveizer, *Vblizi Tolstogo*, p. 270.

Mechnikov and other detractors never explained how it was that Kovalevskaia was able to achieve such outstanding results in areas in which Weierstrass never worked. But the mud from their accusations stuck. Litvinova, the Kazan mathematician A. V. Vasilev, the Moscow mathematician D. A. Tarasov, and others have had to argue against this in great detail.[37]

To some extent, Kovalevskaia herself fueled notions of her excessive dependence upon Weierstrass. She always referred to him as the originator of her ideas, and quoted him extensively. Even when she developed solutions to problems which he had mentioned only in the vaguest way, Sofia described the work in her papers as already-completed results Weierstrass had given her. In this way, Kovalevskaia tried to compensate Weierstrass, who shared with her his hurt at not being cited more often by his students.[38] Sofia took excessive care to avoid wounding him, even at the expense of her own reputation for originality.

Typically, a research mathematician begins her dissertation work under the close supervision of an adviser, and gradually, during the early stages of her career, progresses toward independence. This does not mean that the young mathematician is necessarily drifting very far away from the research interests of her adviser. Rather, the adviser is no longer guiding her student toward fruitful problems, no longer supplying key ideas to aid the investigation.

Kovalevskaia's relationship with her adviser followed the form outlined above. Her three dissertations were on problems suggested to her by Weierstrass, who told her which of his methods would probably yield the best results. But she showed considerable independence even at this stage. Weierstrass wrote to the mathematician Paul DuBois-Reymond that he had expected entirely different, less important results than what Sofia discovered in her research on partial differential equations. He only gave her the problem. She worked out her own

---

[37]See Litvinova, p. 90; A. V. Vasilev, "Pamiati S. V. Kovalevskoi," *Rech'*, 29 January 1916; D. A. Tarasov, "Professor S. V. Kovalevskaia," *Russkie vedomosti*, 30 January 1901; Polubarinova-Kochina, "Zhizn'," *Pamiati*, pp. 63–4.

[38]Weierstrass, 1 January 1875 letter to Kovalevskaia, and others in *Briefe/ Pis'ma*, p. 57ff. and *passim*.

251

solution along different lines than he had envisioned, Weierstrass reported with satisfaction.[39]

Kovalevskaia's later work, after she returned to mathematics in 1881–82, was independent from Weierstrass. To be sure, she used his methods of analysis and function theory, and it was he who suggested the light refraction topic. In her autobiographical sketch, she said that her studies with him had "finally and irrevocably defined the direction I would follow in future scientific work: all my articles are done in the spirit of Weierstrass' ideas."[40] However, in her mature mathematical papers, she applied those ideas to areas in which Weierstrass had little interest. This is what constituted her originality.

According to Mittag-Leffler, Weierstrass thought that Kovalevskaia was the most gifted of all his disciples.[41] When one considers that his students included Leo Königsberger, Georg Frobenius, Hermann Schwarz, Carl Runge, Mittag-Leffler, and other prominent mathematicians, it becomes clear that Weierstrass valued Kovalevskaia's abilities extraordinarily highly.[42] It is perhaps unnecessary to point out that "most gifted" is not equivalent to "best." Rather, Weierstrass felt that Kovalevskaia had the greatest potential of his students as of 1876, when he made the comment.

Hermite, Picard, Poincaré, Bertrand, Weierstrass, Schwarz, Hensel, Kronecker, and the other great and near-great mathematicians of late nineteenth-century Europe accepted Kovalevskaia fully and admitted her into their ranks. This cannot be emphasized too much. Sofia was not seen as some kind of freak, clinging precariously to the fringes of the mathematical community. Rather, she was viewed as a respected colleague and an equal.

---

[39]15 December 1874 letter in "Briefe von K. Weierstrass an Paul DuBois-Reymond," *Acta Mathematica*, 39 (1923), pp. 204–05.

[40]"Avtobiograficheskii rasskaz," *VP 1974*, p. 371.

[41]Mittag-Leffler, 10 February 1876 letter to Malmsten, in "Weierstrass et Sonja Kowalewsky," p. 172.

[42]See Behnke, "Karl Weierstrass und seine Schule," in *Festschrift* pp. 13–40, for a list of Weierstrass' students, disciples, and people who occasionally attended his lectures.

As Litvinova and Nekrasov remarked, Kovalevskaia did not bring about any revolution in mathematics. None of her works broke entirely new ground; she founded no new school of analysis or mechanics. The Cauchy-Kovalevskaia Theorem is basic to the theory of partial differential equations, however, and has been and will be included in texts on the subject in the foreseeable future. Moreover, her work on the revolution of a solid body about a fixed point has inspired a great deal of research on the problem by later mathematicians.[43]

In addition, it must be noted that Kovalevskaia had excellent mathematical taste, and an instinctive feel for which new ideas in mathematics would yield the biggest dividends in years to come. Sofia was always aware of new developments and directions in research. Her early sponsorship of Weierstrassian analysis was only one of the instances of this.

Her colleagues were sometimes amazed by her immediate grasp of important modern trends in research. Mittag-Leffler, for example, was moved to study Henri Poincaré's most recent work under Kovalevskaia's direction. Gösta did not understand it, he wrote to Weierstrass, but Sofia was so enthusiastic about it that it must be significant. Charles Hermite felt similarly about Sofia's espousal of Dirichlet.[44]

This superior taste and grasp of the frontiers of the mathematical research of her time led Kovalevskaia to seek ways of applying the techniques of one area to the problems of another. Hence her interest in Dirichlet's number theoretic work, although she herself was an analyst. Hence also her use of abelian functions in astronomy and mechanics.

Kovalevskaia was so enthusiastic about Weierstrass' function theory that she was convinced it could be applied to the resolution of many problems in mathematical physics and mechanics. She tried to persuade her skeptical colleagues, but often was

---

[43]See *Istoriia otechestvennoi matematiki* (Kiev: Naukova Dumka, 1967), vol II, pp. 235, 290–92, 350, 361, 451, 488; "O znachenii nauchnykh rabot S. V. Kovalevskoi," *VIP 1951*, pp. 537–42; [Polubarinova] Kochina, *Nikolai Evgrafovich Kochin* (Moscow: Nauka, 1979), pp. 129, 142; Lev Gumilevskii, *Chaplygin* (Moscow: Molodaia Gvardiia, 1969), pp. 44, 47; K. E. Tsiolkovskii, "Iz moei zhizni," *Izvestia* and *Pravda*, 20 September 1939; Roger Cooke's forthcoming *The Mathematics of S. V. Kovalevskaya*.

[44]Mittag-Leffler, 15 May 1888 letter to Weierstrass; Hermite, 26 February 1891 letter to Mittag-Leffler; both in the IML.

unsuccessful. They seem to have thought that she was letting her reverence for Weierstrass carry her away.

Ninety years after the fact, however, Kovalevskaia's instincts have been proved correct in at least one major area of current research on partial differential equations. The following quotation from the Introduction to the 1976 mathematical paper "Nonlinear Equations of the Korteweg-DeVries Type" seems a fitting way to close this discussion of Kovalevskaia's mathematics:

We present here an excerpt from a letter written by S. V. Kovalevskaia in December 1886 [to Mittag-Leffler]:

"He (Picard) reacted with great skepticism when I told him that functions of the type

$$y = \frac{\theta(cx + a, c_1x + a_1)}{\theta_1(cx + a, c_1x + a_1)}$$

can be useful for integrating several differential equations . . ."

The analysis done by the authors [of this article] has proved that Picard's doubts were justified only during the ninety years which passed between S. V. Kovalevskaia's works and 1974 work on K[orteweg] d[e] V[ries] equations. They are justified no longer.[45]

Not every mathematician has the posthumous compliment of having her speculations confirmed ninety years after disbelieving colleagues had discounted them. Kovalevskaia's ideas are receiving more attention in certain mathematical circles than they have at any time since she was awarded the Prix Bordin in 1888. The Kovalevskaia top and her asymptotic method are now being reexamined by mathematicians in the United States and the Soviet Union, who have shown how Kovalevskaia's work ties in with current themes in their re-

[45]B. A. Dubrovin, V. B. Matveev, S. P. Novikov, "Nelineinye uravneniia tipa Kortevega-deFriza," *Uspekhi matematicheskikh nauk*, 31, No. 1 (187) (1976), p. 61. I am indebted to Professor Kenneth Goodearl of the University of Utah Department of Mathematics for calling my attention to this article.

search.[46] They are also sifting through Sofia's correspondence with Mittag-Leffler, Picard, and others, with the idea that they will find other connections to modern mathematical theory.

Such renewed interest in the work of a mathematician so long dead is unusual. It is more common for the names of mathematicians prominent in their own time to be forgotten as their work is superseded by newer results. That Kovalevskaia is to have a second period of recognition is indicative of the depth of her mathematical thought.

Indeed, several mathematicians have said that until recently, they had placed Kovalevskaia in the ranks of the minor luminaries of the nineteenth century. But now, with mathematical physicists and specialists in theoretical mechanics reevaluating her work, the respect in which Kovalevskaia is held by experts has been enhanced.[47] Now more than ever it appears that Kovalevskaia deserves the title of the greatest known woman scientist before Maria Sklodowska Curie.

[46]M. Adler and P. Van Moerbeke, "Kowalewski's Asymptotic Method, Kac-Moody Lie Algebras and Regularization," *Communications in Mathematical Physics*, 83, No. 1 (1982) pp. 83–106; Mutiara Buys, *The Kovalevskaya Top* (Diss. New York University, 1982).

[47]Conversations with Professors Mutiara Buys, A. T. Fomenko, Mark Kac, Victor Kac, and Iu. I. Manin.

fourteen

KOVALEVSKAIA'S LITERARY WORKS

Sofia Kovalevskaia wrote a number of plays, novellas, poems, essays, and sketches, many of which were unfinished at the time of her death in 1891. With the exception of her *Memories of Childhood*, however, most of her works are of marginal literary interest at the present time. Sofia's style is uneven, her characterizations are inconsistent, and her narrative point of view too naive and sentimental to appeal to modern tastes.

But at the time of their first appearance in Russia, in the late 1880s and early 1890s, Kovalevskaia's writings were read with eagerness. Some reviewers liked her style, but for the most part they concentrated their attention on the political content of Sofia's works. A. N. Pypin of *Vestnik Evropy* (European Messenger), for example, commented that "however few her works, they constantly reflect the motives of social life."[1] By this, he meant to say that Sofia had progressive beliefs, and that compared to the frothy, trivial productions of most other belle-lettrists, her writings were worth reading.

Kovalevskaia's works appealed to the liberal and radical intelligentsia in Russia, and to those segments of the reading public who cherished hopes for reform of the tsarist government. Her readers ranged politically from the famous communist-feminist Aleksandra Kollontai to at least one member of Tsar Nicholas II's own family.[2]

[1] A. V. [A. N. Pypin], "Literaturnye sochineniia S. V. Kovalevskoi," *Vestnik Evropy*, No. 1 (1893), p. 892.

[2] T. L. Shchepkina-Kupernik, "O pervom predstavlenii dramy 'Bor'ba za

Part of the wide acceptability of Kovalevskaia's writings resulted from the ambiguous way in which their messages were phrased. Her vague language satisfied the tsarist censors, and more than satisfied the public. People on the moderate-to-liberal side of the debate who read Sofia's writings could declare that she had meant nothing more than precisely what she had written: a mildly expressed hope for nonviolent change in the distant future. And people of radical-to-revolutionary leanings could read between the lines of her works and assert that secretly, Kovalevskaia was one of them.

In reality, Kovalevskaia was both intellectually and emotionally closer to the revolutionaries than she was to the moderates. A participant in the social and political ferment of her day, she prided herself on her contacts with socialists and her reputation as a radical. She took special care, however, to avoid offending the sensibilities of the tsarist censors.[3]

Kovalevskaia never tried to repudiate her connections with conspiratorial circles. On the contrary, she felt guilty because she could not bring herself to make the sacrifices that her sister Aniuta and her friends Maria Jankowska-Mendelson, Petr Lavrov, Georg Vollmar, S. A. Padlewski, Natalia Armfeldt and others had made. Although she did far more than the vast majority of her contemporaries, she was always uncomfortably conscious of having contributed less than she could have. She preserved her contact with émigré revolutionary circles because she admired and respected their dedication, and wished to help them in whatever small ways she could.

In a sense, Kovalevskaia atoned for putting her career as a mathematician first by writing novels and plays with political themes. Virtually everything she produced in her six-year career as an author—essays and reminiscences included—touched upon at least one contemporary question: rebellion against tradition, the deceptions of pseudoscience, public education, feminism, socialism, communism. Her views were sometimes romantic and idealistic, but it was always clear that

schast'e,'" *Pamtati*, p. 143; A. N. Pypin, undated [1890] note to Kovalevskaia, IML.

[3]See drafts of small sections of *Nigilistka* and a sketch meant for *Bor'ba za schast'e* in the Kovalevskaia papers of the IML. In these, overtly political phrases have been crossed out in favor of more vague terminology.

Kovalevskaia intended her work to be taken as a political statement.

A case in point is Kovalevskaia's and Anna Carlotta Leffler's *The Struggle for Happiness: How It Was* and *How It Might Have Been* (1887).[4] The characters are poorly drawn, the speeches are stilted, and inconsistencies arise because the play was conceived by Sofia as a purely Russian story and then transferred by Anna Carlotta to a Swedish locale. The play is really two companion pieces with the same characters and prologue, but with the action taking divergent paths.

In *How It Was* the noble heroine Alisa turns her back on her impoverished suitor Karl and, out of a misguided sense of duty, marries her cousin Hjalmar. Hjalmar, in turn, resists his attraction to Karl's sister Paula in order to marry Alisa and in that way obtain her inheritance. Everyone is unhappy. Karl marries another woman, Hjalmar commits suicide at the end, and all the main characters come to realize that their misery is the result of their own selfishness, misplaced pride, and greed.

In *How It Might Have Been*, the characters begin by complicating their lives in many of the same ways as in *How It Was*. Alisa and Hjalmar marry, Paula goes away to Stockholm, Karl becomes bitter and withdrawn. But in *How It Might Have Been*, Alisa, Karl, Hjalmar and Paula eventually find the courage and moral strength to rectify the mistakes of their lives. Alisa and Hjalmar divorce, and Karl and Alisa marry. Hjalmar renounces his claim to Alisa's inheritance, and finally declares himself to the low-born Paula.

Moreover, Karl and Alisa are sincerely interested in the good of the workers at Alisa's factory. The play includes much discussion of education and equality for the common people, and several debates between Alisa and her conservative aunt on the worth of the workers. *How It Might Have Been* ends with Karl and Alisa forming a Chernyshevskii-like cooperative in their factory, where they will share equally with all their employees.

Neither *How It Was* nor *How It Might Have Been* were popular in Sweden. They were too alien in both form and content to be acceptable to the Swedes. Scandinavian audiences were unin-

---

[4]First published in Swedish as *Kampen för Lyckan* in 1887, and in Russian as *Bor'ba za schast'e* in 1892.

terested, for the most part, in the utopian socialism of *The Struggle for Happiness*. Moreover, the psychological theme of rebellion against convention had been treated more capably in other works by Henrik Ibsen and even Anna Carlotta Leffler—when she was not working in conjunction with Sofia.

By contrast, in Russia *The Struggle for Happiness* had some success. The actress Lydia Yavorskaia (Princess Bariatinskaia) chose *How It Might Have Been* for her benefit performance at the Korsh Theater in Moscow in 1895.[5] The play received favorable reviews which tended to praise the wisdom of Yavorskaia's choice of drama and the politics of the play more than the talent of the actors or the literary merit of Kovalevskaia's brainchild.[6]

Whenever key words like "workers," "union," "happiness for the majority" were uttered, the audience, composed mostly of students and young workers, would burst into thunderous applause. Yavorskaia and the other actors sometimes had to wait several minutes for the noise to stop after Alisa's stirring declarations of solidarity with workers. In fact, one reviewer noted that although the play on the whole was tedious, "viewers were obviously carried away by many of its scenes."[7]

The Struggle for Happiness was not the most political of Kovalevskaia's works, but it was among the most political of those which were approved for publication in Russia. *A Nihilist Girl*, also titled *Vera Vorontsova* or *Vera Barantsova* in a futile attempt to deceive the censors as to the nature of the novel, was repeatedly banned in Russia. It first appeared in 1892 in Swedish and in a Russian edition printed in Geneva by Maksim Kovalevskii.[8] But it was struck from the first Russian edition of Kovalevskaia's collected works in 1893. Even a line in Maksim's

[5]E. Halpérine-Kaminsky, "Madame Lydie Yavorskaïa. Princesse Bariatinsky," *Le Théatre*, No. 9 (1902), pp. 19–20.

[6]See reviews by "V" and "Sigma" [S. N. Syromiatnikov] in the 20 and 21 April 1895 *Novoe vremia*, and the memorial article by N. E. Efros, "S. V. Kovalevskaia—dramaturg," *Russkie vedomosti*, 29 January 1916.

[7]Shchepkina-Kupernik, p. 142; "V", *Novoe vremia*, 20 April 1895.

[8]*Vera Vorontzoff* (Stockholm: Albert Bonniers, 1892); *Nigilistka* (Geneva: Volnaia Russkaia Tipografiia, 1892).

Introduction to the legal Russian edition of the 1893 Collected Works had to be removed because it mentioned *A Nihilist Girl*.[9]

A Nihilist Girl appeared only twice in Russia before the Revolution of 1917. Once was in 1906, in the relatively tolerant atmosphere engendered by the Revolution of 1905. Then, in 1908, the novel was published in Prague in Czech translation. The translation was allowed into Russia, because the censor believed "a book in the Czech language will be inaccessible to the Russian [reading] public."[10]

Despite the government's attempts to ban *A Nihilist Girl*, the Russian intelligentsia was familiar with the book from its first appearance in 1892. German, French, Polish, and English translations were smuggled into St. Petersburg from abroad, as was Maksim Kovalevskii's Geneva edition. The plot and politics of *A Nihilist Girl* were open secrets. Litvinova, Pypin, Maksim Kovalevskii and others alluded to the novel in their writings on Kovalevskaia, and it is clear that they expected their readers to know exactly what their allusions meant.[11] As a result of *A Nihilist Girl*, Sofia's reputation among the radical and liberal intelligentsia was enhanced even further.

A Nihilist Girl is the story of a young woman, Vera Vorontsova, who sacrifices her own personal happiness to marry a condemned revolutionary. According to tsarist practice, by marrying him she could probably cause his sentence to be reduced. The novella gives a sympathetic portrayal of the young people of the "go to the people" movement who were arrested in 1874. It also contains autobiographical material concerning Kovalevskaia's own return to St. Petersburg in 1874. The last scene has Sofia crying as she sees Vera off to Siberia to be with her exiled husband. Vera's final words are: "You are crying so

[9]*VP 1974*, pp. 518–19, n. 1, and p. 544. The IML has a copy of the 1893 Collected Works with the sentence mentioning *Nigilistka* still in the Introduction.

[10]*VP 1974*, 518–20; protocols of the 20 August 1908 censors' meeting, p. 520.

[11]See Litvinova, p. 92; Pypin, pp. 890–91; M. Kovalevskii, Introduction to Kovalevskaia's *Literaturnye sochineniia* (St. Petersburg: Stasiulevich, 1893), p. v; Sergius Stepniak and William Westall, eds., *Vera Barantzova* (London: Ward & Downey, 1895), pp. ix–x.

hard for me? Ah, if you only knew how sorry I am for all of you who remain behind!"[12]

With the exception of *Memories of Childhood*, *A Nihilist Girl* is the most beautifully crafted of Kovalevskaia's works. Sofia was at her best when writing fictionalized accounts of events through which she had lived. For this reason, both *Memories of Childhood* and *A Nihilist Girl* are better written and more believable than *The Struggle for Happiness*.

In the two former works, Kovalevskaia chose to embroider reality in order to present a specific picture: Aniuta as the center of Sofia's universe, Vera Vorontsova as the glorious martyr-heroine. But the fictionalized elements in *Memories of Childhood* and *A Nihilist Girl* were under tighter control than in *The Struggle for Happiness* because they were fitted onto a framework of actual experience.

A Nihilist Girl was based on real happenings. Vera Vorontsova was in fact Vera Goncharova, a noblewoman and distant relative of the poet Pushkin's wife. Goncharova came to Kovalevskaia to ask her to find a way for Vera to aid the arrested populists. One of the populists' lawyers told Vera and Sofia that the best way to do that would be to marry one of the prisoners. In real life, Kovalevskaia used her acquaintance with Dostoevskii to help Goncharova meet the political prisoner I. Ia. Pavlovskii.[13] In the novella, however, Sofia obscures her own role and gives credit for all initiative to Vera.

Goncharova's story was not as simple and happy as the fictional version. Pavlovskii escaped from Siberia and lived with her for a time in Paris, but he was cruel and brutal. Kovalevskaia sought out Goncharova in Paris in 1882. When she saw how miserable Vera was, Sofia gave her money and her own passport to return to Russia. Later, Pavlovskii threatened to throw acid in Sofia's face if she did not reveal Vera's Russian address. But Kovalevskaia called his bluff, and he left her apartment without doing anything.[14]

[12]*Nigilistka*, VP 1974, p. 156.

[13]Undated 1876 letter from Kovalevskaia to Dostoevskii, *VIP 1951*, p. 247.

[14]Mendelson, pp. 174–75; Kovalevskaia, 20 February 1882 diary entry, IML.

Maria Jankowska-Mendelson, who witnessed some of this episode, once asked Sofia why she did not write the true story. Kovalevskaia replied: "You see, in the West our [Russian] truth does not sound plausible and would impress the reader as a sick fantasy."[15] An added reason for Sofia's reticence was that she intended her novella as a statement of her political beliefs. The sordid reality of that particular incident would have clouded the political issue.

Like *Memories of Childhood*, *A Nihilist Girl* is of considerable historical interest. Besides the sympathetic portrayal of the radical movement of the 1870s, the novella contains a description of a girl's childhood and adolescence in a noble household around the time of the emancipation of the serfs in 1861. Although ostensibly the girl is Vera, much of the picture fits the young Sofa equally well. Through *A Nihilist Girl*, therefore, Kovalevskaia tells more of her early childhood, and further explains the development of her political ideas.

The Nihilist (1890?), which was unfinished at the time of Kovalevskaia's death, was also a political statement. The novel was to be a fictionalized biography of the radical publicist Nikolai Chernyshevskii and his wife Olga. The portion Sofia finished before she died contains fascinating descriptions of the political and intellectual circles of the 1860s. There are brief sketches of the leading literary figures of the early 1860s, and portraits of the new women students—Nadezhda Suslova, Natalia and Maria Korsini, and others. The manuscript breaks off mid-sentence. According to Ellen Key, who was with Sofia during her last illness, Kovalevskaia meant to end the novel with a midnight knock on Chernyshevskii's door, signifying his arrest.[16]

The main body of *Vae Victis* (Woe to the Vanquished), a projected novel set on the Riviera, a novella entitled *At the Exhibition*, and a novel *Husbands and Wives* are only some of the

[15]Mendelson, p. 175.

[16]Leffler, "Sonya Kovalevsky," pp. 290–91; *The Nihilist* appeared for the first time as *Nigilist*, VP 1974, pp. 157–81.

literary projects Kovalevskaia left unfinished.[17] They are all competently written, and of at least some historical interest because of their autobiographical elements. For example, the heroine of *Vae Victis* is a *gymnasium* teacher of mathematics, the hero of the Riviera novel is a thinly disguised Maksim Kovalevskii, and Sofia's cousin Mikhail was to figure in *At the Exhibition*. But these and Kovalevskaia's other novellas were broken off at an early stage. For most of them even the main outline of the plot remains unclear.

Kovalevskaia's poetry has survived in more complete form. But like her novellas, Sofia's verse is more interesting for its content than for its style. One poem in particular, "A Husband's Lament," is notable for its half-joking diatribe against educated wives.[18] It is written from the point of view of the man, who complains that women's notions of equality have hurt men more than they have helped women. The narrator warns that men who try to flirt with women by humoring the women's ideas of equality risk being taken more seriously than they had intended. The poem is an amusing piece of nonsense, possibly written to tease or goad Vladimir Kovalevskii.

In the last six years of her life, Kovalevskaia wrote a series of essays for Russian and European journals and newspapers. The pieces were popular. Sofia's terse, matter-of-fact narrative style was suited to the essay format. As a result, editors in Western Europe, Russia, and even America encouraged her to continue her efforts in this area.[19]

Kovalevskaia's "Memories of George Eliot," "M. E. Saltykov (Shchedrin)," and "Three Days in a Peasant University in Sweden," were mentioned above. Sofia also wrote two articles

[17]The 1892 Swedish *Vera Vorontzoff* contained several of Kovalevskaia's unfinished works, among them the prologue to *Vae Victis*. *Vae Victis* appeared first in Russian in the April 1892 issue of *Severnyi vestnik*, pp. 237–45. All of Kovalevskaia's collected scraps except for *Husbands and Wives* appear in *VP 1974*. I found *Husbands and Wives* in the IML; it has never been published.

[18]"Zhaloba muzha," *VP 1974*, pp. 319–20. This poem was published for the first time in 1974.

[19]Letters from Theodore Stanton and Martha Yoote Crowe (Americans), A. N. Pypin, B. B. Glinskii, Santiago Barberena (El Salvador!) and others soliciting manuscripts, Kovalevskaia papers, IML.

about hospitals for the poor in Paris.[20] These articles are especially interesting because of Kovalevskaia's skepticism toward one of the prevailing pseudoscientific crazes of her day—hypnotism. Kovalevskaia visited the wards of two Paris physicians who claimed to be able to heal patients by means of hypnotism. Their patients were mostly young, impoverished women who agreed to be guinea pigs in the doctors' experiments so that their other ailments would be treated free of charge.

It is apparent throughout the two articles that Kovalevskaia was suspicious of the doctors' methods and claimed results, and was furious at the indignities suffered by the helpless women patients. Her attack is subtle, however. She does not deny that there are perhaps some doctors legitimately and seriously exploring hypnotism as a tool in medical practice. Nor does she accuse the two Parisian doctors of outright charlatanism. Rather, she calmly recounts her experiences in the hospital and cleverly leaves the reader to draw her own conclusions.

All of Kovalevskaia's novellas, plays, and essays contributed to her posthumous reputation as a writer. But by far the most popular and enduring of her works is the lyrical, evocative, beautifully written *Memories of Childhood* (1888–89). Like any memoirs of childhood years, they are heavily fictionalized. As the Russian critic A. Volynskii pointed out: "It is not possible *to remember* one's youthful feelings. Describing childhood and adolescence, one necessarily has to re-create or, if you prefer, simply *create*."[21]

Although discerning critics at the time recognized the fictional element in Kovalevskaia's memoirs, on the whole they found them so charming and intellectually engaging that no one chose to quibble with her facts. Virtually all Russian reviewers raved about *Memories of Childhood*.[22] Western European

[20]"V bol'nitse 'La Charité' " and "V bol'nitse 'La Salpêtrière' " in *Russkie vedomosti*, 28 October and 1 November 1888; both in *VP 1974*, pp. 267–81.

[21]A. Volynskii, "Literaturnye zametki," *Severnyi vestnik*, No. 10 (1890), p. 156.

[22]See *ibid.*, *passim*, and the review's continuation in No. 3 (1891), pp. 147–

critics were pleased as well. One London journalist even went so far as to compare Kovalevskaia's memoirs to Turgenev's great novel *Fathers and Sons*.[23]

Memories of Childhood gives a lively portrait of a provincial gentry family in the late 1850s and early 1860s. Western European readers were fascinated by the glimpse into Russian family life. Russian readers were also interested in the wealth of domestic detail. But as usual, they were drawn to the politics of the memoirs as well. Kovalevskaia traces the political development of her sister Aniuta and herself in several chapters: "My Uncle Petr Vasilevich," "My Sister," "Aniuta's Nihilism," and the initially-suppressed "Memories from the Time of the Polish Rebellion."[24]

Unfortunately, Kovalevskaia stopped her reminiscences just when they were becoming most interesting—around the time of her fifteenth birthday. She had intended to carry them further, at least through her student days in Heidelberg and Berlin.[25] But she died before she could complete the project.

Kovalevskaia's early death shocked her contemporaries. Since she died just as her first full-length literary works were appearing, reviewers possibly assigned more importance to her talents as a writer than they would otherwise have done. Moreover, Sofia was already a celebrity in such an entirely different field as mathematics. There was curiosity value in all of her literary productions, regardless of their artistic merit. Finally, Kovalevskaia's political opinions were congenial to many of the intellectuals of her time.

It is therefore no wonder that Kovalevskaia's works were praised so enthusiastically in the years after her death. According to Stepniak-Kravchinskii, a Russian revolutionary living in exile in London:

63; "Bibliograficheskii otdel': Periodicheskie izdaniia," *Russkaia mysl'*, No. 8 (1890), pp. 390–93; Pypin, pp. 889–94.

[23]T. P., "A Book of the Week. Sophie Kovalesky [sic]," *The Weekly Sun* (London), 23 September 1894. (The review continues in the 30 September issue.)

[24]This appeared in the 1892 Swedish *Vera Vorontzoff*, but was not published in Russia until *VP* 1974. For more details, see p. 51, n. 45 above.

[25]Kovalevskaia, 29 September 1890 letter to A. N. Pypin, *VIP 1951*, p. 310.

None of our great writers has been more generally admired or more sincerely mourned. After her death Russian literature was flooded with articles on her life, her personality, and her work, both as a scientist and authoress.

Very soon, the strong liberal, or rather radical opinions which she had held became known. Moreover, was not all her activity an everliving protest against the yoke of convention and tradition? Her name became a reanimating watchword for the liberal party, and an expression of sympathy with her work a declaration of liberal aspirations.

So roundabout a way of proclaiming opinions may appear strange to English people. But so it was, and to such an extent that the Government deemed it expedient to issue a secret order to the press forbidding any further mention of Mme. Kovalevsky's name.[26]

That the tsarist government banned further mention of Kovalevskaia's name in the Russian press seems to have been speculation on Stepniak's part. At least, no such secret order has been found. However, there does seem to have been mild exasperation among some officials at the adulation Kovalevskaia was receiving. Minister of the Interior Durnovo is said to have opposed a translation of Anna Carlotta Leffler's memoirs of Kovalevskaia with the words: "People have already concerned themselves too much with a woman who, in the last analysis, was a nihilist."[27]

[26]Stepniak, Introduction to *Vera Barantzova*, pp. ix–x.

[27]27 January 1893 letter from Theresa Gyldén to Iulia Lermontova, in Polubarinova-Kochina, "Zhizn'," *Pamiati*, p. 61.

fifteen

CONCLUSION

It is unfortunately the custom among some feminist writers today to exaggerate the past achievements of women, to claim that this or that famous woman was the equal of or better than the greatest males in the same field. Since it is often impossible to prove these claims, as in many cases they are not true, the net effect of their efforts is to belittle the real, important achievements of the women they wish to aggrandize.

On the other hand, it must be admitted that inflated claims of women's achievements are a natural reaction to past silence about or denigration of their feats. For example, the mathematician Sophie Germain's name does not appear on the Eiffel Tower among the list of winners of the Grand Prix of the French Academy of Sciences, although she won the prize in 1816. This omission is particularly glaring since her work on the elasticity of metals was partly responsible for making the construction of the Tower itself possible.[1]

In the past, Kovalevskaia has often been ignored or disparaged by historians.[2] More recently, however, her achievements have been obscured by exaggerated claims and what amounts almost to idolatry in some accounts.[3] This is a pity,

[1]Mozans, *Woman in Science*, p. 156.

[2]See, for example, E. T. Bell's *Men of Mathematics*, pp. 406–32; Hermann Weyl, "Emmy Noether," *Scripta Mathematica*, 7, No. 3 (July 1935) pp. 219–20, for disparaging evaluations of Kovalevskaia.

[3]See, for example, Lynn Osen, *Women in Mathematics*; Eileen L. Poiani, "The Real Energy Crisis," in *Mathematics Tomorrow* (New York: Springer, 1981), pp. 155–63; Edith H. Luchins and Abraham S. Luchins, "Female Math-

because her accomplishments need no embellishment. She was the first woman in modern Europe to receive a doctorate in mathematics and the first woman in Europe outside of Italy to have a university chair in any field. She was the first woman to be elected to the Russian Imperial Academy of Sciences. She won the Prix Bordin of the French Academy of Sciences. She was the first woman to become an editor of a major scientific journal (*Acta Mathematica*). Hers is one of the three classic cases of revolution of a solid body about a fixed point, and her work on partial differential equations is basic to that field.

This dry listing of Kovalevskaia's accomplishments does not begin to tell the whole story of her exciting, adventurous life, however. Kovalevskaia was a daughter of the 1860s, a champion of women's rights, and a political progressive who remained true to the ideals of her youth to the end of her days. She, her sister Aniuta, and their friends Anna Evreinova, Iulia Lermontova, Elizaveta Litvinova, Nadezhda Suslova, Maria Bokova-Sechenova—all fought for personal independence and the right to a university education. They overcame the obstacles set in their way by their society and, with the exception of Aniuta, all became professionals of some kind.

Sofia Kovalevskaia was a phenomenon in her own time, a symbol of what women could achieve under the right circumstances. Women's rights advocates pointed to her with pride, and she herself was always conscious of her status as a representative of the new woman.

Kovalevskaia never rejected the women's movement, never acted as if her success were due only to her own abilities. She acknowledged her debt to the women's movement of her day and to the socio-political creed of the Russian youth of the 1860s. She knew that without the supportive, unprejudiced, free atmosphere that attended nihilist circles in her youth, she and her friends would probably never have broken away from the traditionalism of patriarchal Russian society.

Kovalevskaia, Litvinova, Lermontova, Suslova, Bokova-Sechenova and Evreinova were among the first women in modern times to receive advanced degrees in Europe. These

---

ematicians: A Contemporary Appraisal," in *Women and the Mathematical Mystique* (Baltimore: Johns Hopkins University Press, 1980), p. 14.

women, coming from backward Russia, approached the European universities with determination, and persuaded them to grant degrees. There had been some auditors before them, but Kovalevskaia and her friends were the first to receive official doctorates in medicine, the sciences, and law.[4]

As a scientist, Sofia Kovalevskaia was the most successful woman of her generation. She was a professional mathematician—possibly the first woman since Hypatia of fifth-century Alexandria who could be so called. To the best of our present knowledge, Kovalevskaia was the finest woman scientist anywhere in the world before the twentieth century. It was only in the first decade of this century that another woman, the chemist/physicist Maria Sklodowska Curie, attained prominence in the sciences. Coincidentally, Maria Sklodowska Curie, a Slavic woman of the generation after Kovalevskaia, left Warsaw and the Russian empire to study in Paris in 1891, the year of Sofia's death.

Kovalevskaia numbered as her peers the finest mathematicians of her day. She received her education in some of the best European centers of the time—Berlin and Heidelberg—and therefore had the finest technical training in the world. When Weierstrass called Kovalevskaia his most gifted disciple, it was no ordinary compliment. He was comparing her with the best minds in Western mathematics. All were at European universities, and all were her respectful colleagues.

It should perhaps be noted here that America played little part in the international mathematical community of Kovalevskaia's time. The scientific education received at American colleges and universities in the nineteenth century, and indeed until the 1930s, when so many German and Eastern European scientists fled to the United States, could not compare with that offered at Europe's leading centers. The best mathematics majors at Vassar or even Harvard would have found it difficult to follow Kovalevskaia's most elementary lectures. In fact, few faculty members at American institutions were advanced enough to have succeeded as her dissertation students.[5]

[4]See V. Bëmert, *Universitetskoe obrazovanie zhenshchin* (St. Petersburg: Merkulev, 1873), pp. 7–13.

[5]See David Eugene Smith and Jekuthiel Ginsburg, *A History of Mathematics*

One American mathematician, Fabian Franklin of Johns Hopkins University, said: "Her work was of a far higher grade than any that has as yet been achieved by any American mathematician . . ."[6] For Franklin to make such a statement was praise indeed. A modern expert in partition theory, George Andrews, calls one of Franklin's results "one of the truly remarkable achievements of nineteenth-century American mathematics."[7] That Franklin was so willing to yield precedence to Kovalevskaia shows in what high regard her work was held.

Sofia Kovalevskaia was a celebrity for almost half her life, renowned throughout Europe and America. But she did not allow her fame to change her, to make her proud and aloof from her old friends. She had faults, of course. There were times when she was over-possessive with her daughter, quarrelsome with Maksim Kovalevskii, or too demanding of Gösta Mittag-Leffler's time.[8] Throughout her life, however, she remained a friendly, warm, approachable human being.

Kovalevskaia's contemporaries seemed to sense these qualities in her. From the time of her appointment to Stockholm University in 1883 until her death in 1891, she was bombarded with advice, complaints, and requests from all over Europe and the Americas. Mathematicians submitted manuscripts to her for *Acta Mathematica*, students sent her their work for criticism, and Kronecker and others involved her in their professional feuds.

In addition to technical correspondence, there were frequent

---

in America Before 1900 (Chicago: The Mathematical Association of America, 1934), pp. 102–200 and *passim*.

[6]Fabian Franklin, " 'Masculine Heads' and 'Feminine Hearts.' Apropos of Sonya Kovalevsky," *Century Magazine*, 51 (1895–1896), p. 318. The title is a reference to Isabel Hapgood's sketch of Kovalevskaia in the previous issue of *Century*, in which she characterized the "tragedy" of Sofia's life as the conflict of her mind with her heart. Franklin indignantly responded to Hapgood's denigration of Kovalevskaia and other intellectual women.

[7]George E. Andrews, *The Theory of Partitions* (Reading, Mass.: Addison-Wesley Publishing Company, 1976), p. 9.

[8]See S. Vl. Kovalevskaia's expanded memoirs in [Polubarinova] Kochina, *Sofia Vasilevna Kovalevskaia*, in the appendices, p. 282; Kovalevskaia's 1890 diary entries in Shtraikh, *Sestry*, pp. 330–35; Kovalevskaia's numerous notes to Mittag-Leffler, generally asking him for more time than he could conveniently give, IML.

pleas for Kovalevskaia's photograph, a request for rock samples from Sweden, and a query from a Russian physician about a Swedish gymnast. People from all over tried to trace their relationship to Kovalevskaia, and claim common ancestry. Teachers wrote to her about their students, a mathematics instructress in a girls' school asked whether it was worthwhile to continue teaching girls, and young women asked for advice about whether they should choose a career over marriage.[9]

As a famous mathematician, Kovalevskaia encountered her fair share of the kind of harmless lunatic who often writes to well-known scientists. Numerous people sent her page after page of garbled formulas. One person from Mexico decided that as a woman and recent laureate of the Prix Bordin, she would be sympathetic to his miraculous discoveries. "Whether by accident or skill," he modestly wrote, it had been his fate to arrive at the true method of squaring the circle![10] There was even a curious diatribe written in modern Russian in the Old Church Slavonic handwriting style occasionally used by Russian priests. The pamphlet, addressed "To the rector and faculty of Stockholm University and *in particular* Professor Kovalevskaia," purported to be direct from God, and railed against women.[11]

What is amazing about Kovalevskaia's correspondence is not that she received such letters. She was a celebrity, and celebrities usually receive fan and hate mail. Rather, what is striking is that she kept a large portion of her letters, answered many of them, and even entered into acquaintanceship with some of her correspondents.

Kovalevskaia knew several American women because of their correspondence with her over genealogical matters, or because she had helped them with language difficulties when they traveled in Europe. She spoke English, French, and German, and if she saw someone having problems, would introduce herself and offer to translate. Then she would help

[9]Numerous assorted letters, in French, German, Russian, Polish, and English, in the Kovalevskaia papers of the IML.

[10][Illegible], 14 October 1889 letter to Kovalevskaia, IML.

[11]No date, no name (except God's), but prettily done up in colored inks with a tasseled cord binding, Kovalevskaia papers, IML.

further by giving the tourists letters of introduction to her friends in London, Paris, Berlin and St. Petersburg. Her kindness in performing such small services, and in corresponding with her acquaintances afterwards, made her well-liked by those with whom she came into contact.[12]

Perhaps the most pitiful yet most inspiring of the letters in Kovalevskaia's papers in the Institut Mittag-Leffler is from a Russian woman, Liubov Murakhina. Liubov had never met Kovalevskaia. But like most educated women in Russia, she had been hearing of her since 1870, and had been following Sofia's career.

When Murakhina wrote to Kovalevskaia in the summer of 1889, she was desperate. Like some other women of the sixties, she had married with the idealistic notion of helping her husband to become a better person. He turned out to be a brute and a drunkard, beating her and humiliating her. She bore this treatment for years, but then met her true ideal, who unfortunately was poor and had younger siblings to support. Liubov's brutish husband, after being fired from several positions for drunkenness, demanded monthly maintenance from her or he would take her to court for adultery. She had no money, no hope of employment, and no future happiness without her lover. She turned to Kovalevskaia as a last resort, saying that she would kill herself if Sofia had no advice for her.[13]

Kovalevskaia received Murakhina's letter while she was in France with Maksim Kovalevskii and her daughter, the summer after she was awarded the Prix Bordin. In spite of her preoccupation with Maksim and her activities in French society, she was concerned enough to help Liubov. She induced her publishing acquaintances in Moscow to employ Murakhina as a translator, and told her to visit a friend of Kovalevskaia there for further assistance. Her letter to Liubov was so decisive and practical, and at the same time so sincerely sympathetic to the unfortunate woman's suffering, that Liubov was almost

[12]See letters of thanks from Linda Fulton (1886), and others in the Kovalevskaia papers of the IML.

[13]Liubov Murakhina, undated [late summer 1889] letter to Kovalevskaia, Kovalevskaia papers, IML. The address and date on this were torn off, probably by Kovalevskaia.

pathetically grateful. "I am especially happy," she wrote to Kovalevskaia, "because this time I was not mistaken [in my judgment of people]. I assumed that a woman who has devoted herself entirely to the service of science—and has achieved such brilliant results—ought to be a true *human being* as well."[14]

Perhaps Murakhina, who called herself an "unregenerate idealist," was mistaken about the necessary connection between service to science and kindness to women like herself. In Kovalevskaia's case, however, Liubov's confidence was not misplaced. Sofia Kovalevskaia combined mathematical creativity with concern for others and awareness of social and political issues. Her friends, intimates, and fellow scientists would remember her for her human qualities as well as her mathematical expertise and success as the first woman accepted as a colleague in an all-male profession.

[14]Murakhina, [18 September 1889 postmark] letter to Kovalevskaia, Kovalevskaia papers, IML.

BIBLIOGRAPHY

Bibliographical Note

Of the writings of Sofia Kovalevskaia herself, only two have been recently translated into English: the beautiful *Memories of Childhood*, and a short autobiographical sketch. Both of these appear as *A Russian Childhood*, published by Springer Verlag in the pleasing translation of Beatrice Stillman. But the work does not contain "Memories from the Time of the Polish Rebellion," *A Nihilist Girl*, or "Memories of George Eliot," which are Kovalevskaia's other autobiographical writings.

There are no English versions of Kovalevskaia's mathematical and personal correspondence, her poetry, or her other literary works. For those who are fortunate enough to read Russian, however, all of Kovalevskaia's poetry and literature (except for the fragment *Husbands and Wives*) appears in the 1974 *Vospominaniia**Povesti. Also, some of her personal correspondence is published in the 1951 *Vospominaniia i pis'ma*. Moreover, there is a 1973 bilingual German/Russian edition of Karl Weierstrass' letters to Kovalevskaia (*Pis'ma/Briefe*). These letters span twenty years, and are a lovely testimony to the friendship and cooperation between Kovalevskaia and Weierstrass.

During Kovalevskaia's lifetime and the years immediately following her death, there was a rash of newspaper and journal articles about her life and accomplishments. Friends and relatives wrote personal reminiscences, such as those by Sophie Adelung, Elizaveta Litvinova, F. V. Korvin-Krukovskii, and I. I. Malevich. Even people who barely knew her rushed into

277

print with inaccurate or fabricated accounts; see, for example, the (unintentionally) humorous anonymous article "A Remarkable Russian Woman." There were also several longer works, some of which attempted the comprehensiveness of a biography. The most notable of these were by Kovalevskaia's friends Elizaveta Litvinova, Maria Jankowska-Mendelson, Ellen Key, and Anna Carlotta Leffler.

By far the most widely translated and best known of the early biographical sketches is that of the Swedish writer Anna Carlotta Leffler. Leffler, sister of the mathematician Gösta Mittag-Leffler, was close to Kovalevskaia for two or three years of the latter's professorship in Stockholm. Leffler wrote what purported to be Kovalevskaia's life story as she herself would have wanted it told. In her more candid moments, however, Leffler admitted that what she had written was a "poem," that is, a fictionalized account, based on Kovalevskaia's life:

> From an objective point of view the life-history I have sought to depict of my friend may perhaps be considered not real. But is the objective standpoint the true one when we deal with the interpretation of character?
> Many may contest the justice of my estimate and interpretation; many may judge Sonya's actions and feelings in quite another light, but this in no way concerns me from my point of view. . . . [My biography] will be her own poem about herself as revealed to me. (Leffler, "Sonya Kovalevsky," p. 157.)

Unfortunately, future western commentators chose to ignore the warning contained in Leffler's words. Virtually all biographical sketches of Kovalevskaia in Western Europe and America treat Anna Carlotta Leffler's idiosyncratic and unreliable account as the unbiased truth. Although Anna Carlotta herself did not intend for her "poem" to be the last word on her friend, until the present time many of Kovalevskaia's actions, thoughts, and motivations have been seen through the distorting haze of Leffler's sketch. (See works by Bell, Duarte, Hapgood, Iacobacci, Kramer, Marholm-Hansson, Osen, Perl, Rappaport, Stillman, and Tee.)

Kovalevskaia's mathematical accomplishments have been subjected to similar distortion. Within the mathematical community, the two most frequently cited "authorities" on Kovalevskaia are Eric Temple Bell (*Men of Mathematics*) and Felix Klein (*Development of Mathematics in the 19th Century*). Both of these sources are unreliable. Bell was notoriously inaccurate, and was not one to ruin a good story with facts. Klein was far more responsible, and indeed, praised Kovalevskaia for her "great conceptual talent" and "equally great flexibility." But he also implied that most of her work was probably done by Weierstrass, and he disparaged her later papers. Unfortunately, Bell's and (the negative aspects of) Klein's evaluations of Kovalevskaia's mathematics and her relationship to the mathematical world of her time are generally accepted by most mathematicians and historians of mathematics today. (Exceptions to this are pieces by Karen Rappaport and Roger Cooke.)

Unlike most western work, scholarship in Kovalevskaia's homeland has never been bound by adherence to the accounts of Leffler, Bell, and Klein. Thus, the best research on Kovalevskaia has come from Soviet scholars. P. Ia. Polubarinova-Kochina, S. Ia. Shtraikh, O. A. Oleinik, M. V. Nechkina, and L. A. Vorontsova, in particular, have written excellent articles on areas of Kovalevskaia's life and work, as well as readable, popular biographies. However, little of this work has been translated into English.

For those who wish to read more about Kovalevskaia, the works mentioned above, plus others cited in the footnotes of this book, should prove useful. I recommend the pieces by Cooke, Rappaport, and Polubarinova-Kochina for serious treatment of Kovalevskaia's mathematical work; and my articles "Sofia Kovalevskaia and the Mathematical Community" and "A Few Words on Sofia Kovalevskaya" for descriptions of her professional interactions with mathematicians.

On a more popular level, the works of Anna Carlotta Leffler and Beatrice Stillman should satisfy those who wish for a traditional, romantic portrayal of Kovalevskaia. It should be borne in mind, however, that the picture that emerges from these accounts is distorted by the authors' dislike and ignorance of the mathematical world, and by a love for melodrama. Nevertheless, the sketches make light, entertaining reading, especially if read along with Kovalevskaia's own partly-fictionalized *Memories of Childhood*.

A more serious attempt at a biography is Don H. Kennedy's *Little Sparrow*, which appears as this book goes to press. Kennedy's account is reasonably well-written, detailed, and enjoyable. Unfortunately, it suffers from the fact that Kennedy himself does not read Russian, and so was forced to rely on his wife's translations. More importantly, Kennedy made no use of the archives of the Swedish Academy of Sciences, archival information in the USSR, and many published primary and secondary sources in Russian. The omissions and inaccuracies caused by these gaps in his research become more serious as the book goes on. In later chapters, the errors are considerable. Nevertheless, *Little Sparrow* is an entertaining and informative biography.

ARCHIVES

Tsentral'nyi Gosudarstvennyi Arkhiv Literatury i Iskusstva, Moscow, USSR. (TsGALI)

Institut Mittag-Leffler, Mathematical Institute of the Royal Swedish Academy of Sciences, Djursholm, Sweden. (IML)

KOVALEVSKAIA'S SCIENTIFIC WORKS

"Zur Theorie der partiellen Differentialgleichungen." *Crelle's Journal*, 80 (1875), pp. 1–32.

"Über die Reduction einer bestimmten Klasse von Abel'scher Integrale 3-en Ranges auf elliptische Integrale." *Acta Mathematica*, 4 (1884), pp. 393–414.

"Sur la propagation de la lumière dans un milieu cristallisé." *Comptes rendus de l'Académie des Sciences*, 98 (1884), pp. 356–57.

"Om ljusets fortplanting uti ett kristalliniskt medium." *Of versigt Akademiens Forhandlinger*, 41 (1884), pp. 119–21.

"Über die Brechung des Lichtes in crystallinischen Mitteln." *Acta Mathematica*, 6 (1885), pp. 249–304.

"Zusätse und Bemerkungen zu Laplace's Untersuchung über die Gestalt des Saturnringes." *Astronomische Nachrichten*, 111 (1885), pp. 37–48.

"Sur le problème de la rotation d'un corps solide autour d'un point fixe." *Acta Mathematica*, 12 (1888–1889), pp. 177–232.

"Sur une propriété du système d'equations differentielles qui definit la rotation d'un corps solide autour d'un point fixe." *Acta Mathematica*, 14 (1890), pp. 81–93.

"Mémoire sur un cas particulier du problème de la rotation d'un corps pesant autour d'un point fixe, où l'intégration s'effectue à l'aide de fonctions ultraelliptiques du temps. *Mémoires présentés par divers savants à l'académie des sciences de l'Institut de France*, 31 (1894), pp. 1–62.

"Sur un theoreme de M. Bruns." *Acta Mathematica*, 15 (1891), pp. 45–52.

Nauchnye raboty. Trans. P. Ia. Polubarinova-Kochina. Moscow: AN SSSR, 1948. (Kovalevskaia's collected mathematical papers.)

KOVALEVSKAIA'S LITERARY WORKS
"Avtobiograficheskii rasskaz." *Russkaia starina*, 72, No. 11 (1891), pp. 449–63.
Nigilistka. Moscow: Izd. Kokhmanskogo, 1906.
"Tri dnia v krest'ianskom universitete v Shvetsii." *Severnyi vestnik*, No. 11 (1890), pp. 133–61.
"V bol'nitse 'La Charité'." *Russkie vedomosti*, 28 October 1888.
"V bol'nitse 'La Salpêtrière'." *Russkie vedomosti*, 1 November 1888.
"Vae Victis." *Severnyi vestnik*, No. 4 (1892), pp. 237–45.
Vera Barantzova. Trans. Sergius Stepniak and William Westall. London: Ward & Downey, 1895.
"Vospominaniia detstva." *Vestnik Evropy*, No. 7 (1890), pp. 55–98; No. 8, pp. 584–640.
"Vospominaniia o Dzhorzhe Elliote [sic]." *Russkaia mysl'*, No. 6 (1886), pp. 93–108.
Izbrannye proizvedeniia. Moscow: "Sovetskaia Rossiia", 1982.
Literaturnye sochineniia. St. Petersburg: Tip. M. M. Stasiulevicha, 1893.
Vospominaniia detstva i biograficheskie ocherki. Ed. S. Ia. Shtraikh. Moscow: AN SSSR, 1945.
Vospominaniia i pis'ma. Ed. S. Ia Shtraikh. Moscow: AN SSSR, 1951. 2nd. ed. 1961. (VIP 1951)
Vospominaniia * povesti. Ed. P. Ia. Polubarinova-Kochina. Moscow: Nauka, 1974. (VP 1974)
Kovalevskaya, Sofya. *A Russian Childhood.* Trans. Beatrice Stillman. New York: Springer Verlag, 1978. (Russian Childhood)
Kovalevsky, Sonya. *Sonya Kovalevsky. Her Recollections of Childhood with a biography by Anna Carlotta Leffler, Duchess of Cajanello.* Trans. Isabel F. Hapgood and A. M. Clive Bayley. New York: The Century Company, 1895.
———. *Ur ryska lifvet. Systrarna Rajevski. Öfversättning från förfs manuskript af Walborg Hedberg.* Trans. Walborg Hedberg. Stockholm: z. Heggströms förlagexpedition, 1889.
Note: This it not an exhaustive list of editions of Kovalevskaia's work. Only editions which have been consulted have been included here.

OBITUARY NOTICES
La Bulgarie (Sofia), 12 March 1891.
Istoricheskii vestnik 63 (1891), pp. 900–01.
Kaspii (Baku), 31 January, 2, 5, 9 February 1891.
Knizhki nedeli, No. 7 (1891), pp. 117–24.
Moskovskie illiustrirovannaia gazetka, 1 February 1891.
Moskovskie vedomosti, 4 February 1891.
Novoe vremia, 30 and 31 January, 13 February 1891.
Novosti i birzhevaia gazeta, 31 January, 11 April 1891.
Le Radical Algerien (Algiers), 20 February 1891.
Russkaia mysl', No. 2 (1891), p. 258.
Russkie vedomosti 6, 12, 17 February 1891.
Vestnik Evropy, No. 2 (1891), p. 466.

Obituaries and commemorative articles of particular interest are listed individually in the body of the bibliography.

WORKS CITED

A—va. "Zhenskaia zhizn'." *Otechestvennye zapiski*, No. 3 (1875), pp. 203–58; No. 4, pp. 365–414.

Adelung, Sophie. "Jugenderinnerungen an Sophie Kovalewsky." *Deutsche Rundschau*, 89 (1896), pp. 394–425.

Adler, M. and Van Moerbeke, P. "Kowalewski's Asymptotic Method, Kac-Moody Lie Algebras and Regularization." *Communications in Mathematical Physics*, 83, No. 1 (1982), pp. 83–106.

Aleksandr Mikhailovich Butlerov, po materialam sovremennikov. Moscow: Izdatel'stvo Nauka, 1978.

Ambler, Effie. *Russian Journalism and Politics 1861–1881. The Career of Aleksei S. Suvorin*. Detroit: Wayne State University Press, 1972.

Andreevski, S. "S. V. Kovalevskaia." *Birzhevye vedomosti*, 29 January 1916.

———. "S. V. Kovalevskaia. (Ocherk)." *Novosti i birzhevaia gazeta*, 12 April 1891.

Andrews, George E. *The Theory of Partitions*. Vol. 2 of *Encyclopedia of Mathematics and Its Applications*, Gian-Carlo Rota, ed. Reading, Mass.: Addison-Wesley Publishing Company, 1976.

Angot, E. "Un Peu de Féminisme." *Correspondant*, 10 September 1909, pp. 951–71.

Anuchin, D. N. "K biografii I. I. Mechnikova." *Russkie vedomosti*, 5 July 1916.

———. "V. O. Kovalevskii. (Nekrolog)." *Russkie vedomosti*, 24 April 1883.

———. "V starye gody. (Po povodu stoletnei godovshchiny dnia rozhdeniia Gertsena)." *Russkie vedomosti*, 29 March 1912.

———. "Vladimir Onufrievich Kovalevskii (Nekrolog)." in *Rech' i otchet, chitannye v torzhestvennom sobranii imperatorskogo Moskovskogo universiteta 12go ianvaria 1884 goda*. Moscow: Universitetskaia Tipografiia, 1884.

Appell, Paul. *Henri Poincaré*. Paris: Plon-Nourrit, 1925.

Aptekman, O. V. *Obshchestvo "Zemlia i Volia" 70-kh gg*. Petrograd: "Kolos", 1924.

Ariian, P. N. *Pervyi zhenskii kalendar' na 1899 god*. St. Petersburg: "Trud", 1899. The calendar was published every year until 1912.

Arvor, Camille d'. "Etrangères et Francaises—Les Russes." *Petit Echo de la Mode*, 10 January 1904.

Ashevskii, S. "Russkoe studenchestvo v epokhi 60-kh godov." *Sovremennyi mir*, No. 6 (1907), pp. 12–26; No. 7–8, pp. 19–36; No. 9, pp. 48–85; No. 10, pp. 48–74; No. 11, pp. 108–34.

Atkinson, Dorothy. "Society and the Sexes in the Russian Past." in *Women in Russia*. Ed. Dorothy Atkinson, Alexander Dallin and Gail Warshofsky Lapidus. Stanford: Stanford University Press, 1977, pp. 3–38.

Austin, L. F. "At Random." *The Sketch*, 24 March 1897, p. 390.

B., P. V. "S. V. Kovalevskaia." *Niva*, No. 8, 23 February 1893, pp. 186–87.

Barine, Arvède. "La Rancon de la Gloire.—Sophie Kovalesky [sic]." *Revue des Deux Mondes*, 123 (1894), pp. 348–82.

——. *Portraits de femmes*. Paris: Librairie Hachette et Cie, 1894.

Bell, Eric Temple. *The Development of Mathematics*. 2nd ed. New York: McGraw-Hill Book Company, 1945.

——. *Men of Mathematics*. New York: Simon and Schuster, 1937.

Bëmert, V. *Universitetskoe obrazovanie zhenshchin*. St. Petersburg: Tipografiia P. P. Merkuleva, 1873.

Bessières, A. J. "Le Mouvement Féministe." *Revue Religieuse*, 5 February 1898, pp. 321–25.

Bestuzhevki v riadakh stroitelei sotsializma. Moscow: Izd. "Mysl' ", 1969.

Biograficheskii Slovar' Deiatelei Estestvoznaniia i Tekhniki. Moscow: Bol'shaia Sovetskaia Entsiklopediia, 1958.

Blanc, May Armand. "Trois Étrangères: Sonia Kowalewska, Marie Bashkircheff, Fanny Kemble." *Tribune de la Fronde*, 29, 30, 31 July and 1 August 1893.

Boborykin, P. D. *Vospominaniia*. [Moscow]: Izdatel'stvo khudozhestvennoi literatury, 1965. 2 vol.

Bogdanovich, T. A. *Liubov' liudei shestidesiatykh godov*. Leningrad: Academia, 1929.

Bogucharskii, V. [V. Ia. Iakovlev]. *Aktivnoe narodnichestvo semidesiatykh godov*. 1912; rpt. The Hague: Mouton, 1970.

Bolkhovitinov, V. et al. *Rasskazy o russkom pervenstve*. Moscow: Molodaia Gvardiia, 1950.

Borisiak, A. A. *V. O. Kovalevskii, ego zhizn' i nauchnye trudy*. No. 5 of *Trudy komissii po istorii znanii*. Leningrad: AN SSSR, 1928.

Brandes, Dr. Georg. *Impressions of Russia*. Trans. Samuel C. Eastman. New York: Thomas Y. Crowell, 1889.

Breshko-Breshkovskaia, E. K. "Iz vospominanii (S. A. Leshern, N. A. Armfeld, T. I. Lebedeva, M. K. Krylova, G. M. Gelfman)." *Golos minuvshego*, No. 10–12 (1918), pp. 169–235.

"Briefe von K. Weierstrass an L. Fuchs." *Acta Mathematica*, 39 (1923), pp. 246–56.

"Briefe von K. Weierstrass an L. Königsberger." *Acta Mathematica*, 39 (1923), pp. 226–39.

"Briefe von K. Weierstrass an Paul DuBois-Reymond." *Acta Mathematica*, 39, (1923), pp. 199–225.

Briullov, B. "Vstrecha s F. M. Dostoevskim (so slov P. A. Briullova)." *Nachalo*, No. 2 (1922), pp. 264–65.

Broicher, Charlotte. "Sonia Kovalevsky in Beziehung zur Frauenfrage." *Preuszischen Jahrbücher*, 84 (1896), pp. 1–18.

Broido, Vera. *Apostles Into Terrorists: Women and the Revolutionary Movement in the Russia of Alexander II*. New York: The Viking Press, 1977.

Brower, Daniel R. *Training the Nihilists: Education and Radicalism in Tsarist Russia*. Ithaca: Cornell University Press, 1975.

Bunsen, Marie von. "Sonja Kowalevsky. Eine biographische Skizze." *Illustrierte Deutsche Monatschefte*, No. 82 (1897), pp. 218–32.

Buys, Mutiara. *The Kovalevskaya Top.* Diss. New York University, 1982.
Campbell, Paul J. and Louise S. Grinstein. "Women in Mathematics: A Preliminary Selected Bibliography." *Philosophia Matematica,* 13/14 (1976–1977), pp. 171–203.
Chernyshevsky, N. G. *What is to be Done?* 1863; rpt. New York: Vintage Books, 1961.
"Chestvovanie pamiati S. V. Kovalevskoi." *Russkie vedomosti,* 30 January 1916.
Chukovskii, Kornei. *Iz vospominanii.* Moscow: Sovetskii pisatel', 1959.
Coolidge, Julian L. "Six Female Mathematicians." *Scripta Mathematica,* 17 (1951), pp. 20–31.
Courtney, W. L. "Books of the Day. Sonya Kovalevsky." *The Daily Telegraph* (London), 5 June 1901.
Cross, J. W. *George Eliot's Life as Related in Her Letters and Journals.* Vol. III. Boston: Estes and Lauriat, 1895.
Cruppi, Louise. *Femmes Écrivains d'Aujourd'hui.* Paris: Arthème Fayard, n. d.
———. "La Duchesse de Cajanello (Une Suédoise italienne)." *La Revue* [Revue des Revues], 85, No. 8 (1910), pp. 450–65.
Curie, Eve. *Madame Curie.* Trans. Vincent Sheean. New York: Doubleday, Doran & Company, 1939.
D. I. Mendeleev po vospominaniiami O. E. Ozarovskoi. Moscow: "Federatsiia", 1929.
Danzer, Dr. Klaus. *Robert W. Bunsen und Gustav R. Kirchhoff.* Leipzig: B. G. Teubner Verlagsgesellschaft, 1972.
Davitashvili, L. Sh. *V. O. Kovalevskii.* Moscow: AN SSSR, 1946.
Deherme, G. "L'Esprit du Féminisme." *Cooperàtion des Ideés,* October 1898, pp. 467–72.
Deich, L. G. *Russkaia revoliutsionnaia emigratsiia 70-kh godov.* Petersburg: Gosudarstvennoe izdatel'stvo, 1920.
Derevitskii, A. N. *Zhenskoe obrazovanie v Rossii i zagranitsei.* Odessa: Tip. Isakovich i Beilenson, 1902.
Dictionary of Scientific Biography. C. C. Gillispie, ed. New York: Charles Scribner's Sons, 1973.
Dixon, Marion Hepworth. "Marie Bashkirtseff: A Personal Reminiscence." *Fortnightly Review,* NS 47 (1890), pp. 276–82.
Dom, Edviga. "Nauka i zhenshchiny." Trans. K. V. *Russkaia mysl',* No. 10 (1882), pp. 244–90; No. 11, pp. 318–36; No. 12, pp. 354–84.
Domar, Yngve. "On the foundation of Acta Mathematica." *Acta Mathematica,* 148 (1982), pp. 3–8.
Dostoevsky, Anna. *Dostoevsky: Reminiscences.* Trans. and ed. Beatrice Stillman. New York: Liveright, 1975.
———. *Vospominaniia.* Moscow: "Khudozhestvennaia Literatura", 1971.
Dostoevskaia, L. F. *Dostoevskii v izobrazhenii ego docheri.* Trans. L. Ia. Krukovskaia. Moscow: Gosizdat, 1922.
Dostoevskii, F. M. "Ispoved' Stavrogina." *Byloe,* No. 18 (1922), pp. 227–52.
———. *Pis'ma.* Moscow: Gosizdat, 1928. 4 vol.
Duarte, Francisco Jose. *Biografias* Caracas: Academia de Ciencias, 1969.

Dubbey, J. M. *The Mathematical Work of Charles Babbage.* Cambridge: Cambridge University Press, 1978.

Dubreil-Jacotin, Marie-Louise. "Women Mathematicians." in *Great Currents of Mathematical Thought.* Ed. F. LeLionnais et al. 1962 rpt. New York: Dover Publications, 1971. Vol. I, pp. 268–80.

Dubrovin, B. A., Matveev, V. B., Novikov, S. P. "Nelineinye uraveniia tipa Kortewega de Friza," *Uspekhi mathematicheskikh nauk,* 31, No.1 (187) (1976), pp. 55–136.

Dvizhenie tverdogo tela vokrug nepodvizhnoi tochki. Sbornik, posviashchennyi pamiati S. V. Kovalevskoi. Moscow: AN SSSR, 1940.

Edgren-Leffler, A. Sh. "Alia." *Severnyi vestnik,* No. 3 (1892), pp. 115–59; No. 4, pp. 207–36; No. 5, pp. 79–111 (here, she changes her name to A. K. Leffler, gertsogina di Cajanello); No. 6, pp. 153–94.

Efros, N. E. "S. V. Kovalevskaia—dramaturg." *Russkie vedomosti,* 29 January 1916.

Ehrenreich, Barbara, and Deirdre English. *Complaints and Disorders: The Sexual Politics of Sickness.* Old Westbury, N.Y.: The Feminist Press, 1973.

———. *Witches, Midwives, and Nurses. A History of Women Healers.* Old Westbury, N.Y.: The Feminist Press, 1973.

Eiseley, Loren. *Darwin's Century.* Garden City, N.Y.: Doubleday & Company, 1961.

Elfving, Gustav. *The History of Mathematics in Finland, 1828–1918.* Helsinki: Societas Scientiarum Fennica, 1981.

Emmons, Terence. *The Russian Landed Gentry and the Peasant Emancipation of 1861.* Cambridge: Cambridge University Press, 1968.

Encyclopedia Britanica. 11th ed. Cambridge: Cambridge University Press, 1911.

Engel, Barbara Alpern. "Mothers and Daughters: Family Patterns and the Female Intelligentsia." in *The Family in Imperial Russia.* Ed. David L. Ransel. Urbana: University of Illinois Press, 1978, pp. 44–59.

Entsiklopedicheskii Slovar'. St. Petersburg: I. A. Efron, 1895.

Ernest, John. "Mathematics and Sex." *American Mathematical Monthly,* 83, No. 8 (1976), pp. 595–614.

"Der erste weiblische Professor in Europa." *St. Jalles Stadtes Anzeizer,* 4 March 1891

Eves, Howard W. *Mathematical Circles Revisited.* Boston: Prindle, Weber & Schmidt, 1971.

Evréinoff, A. d'. "Conférence faite à la Solidarité." *Revue des femmes russes et des femmes francaises,* March 1897, pp. 166–70.

Evteeva, Z. A. et al. *Vysshie zhenskie (Bestuzhevskie) kursy: Bibliograficheskii ukazatel'.* Moscow: "Kniga", 1966.

Fahlstedt, Amalia. "Sonja Kovalevskys minne." *Idun,* 26 September 1896.

Fang, J. "Mathematicians, Man or Woman: Exercises in a 'Verstehen Approach.' " *Philosophia Mathematica,* 13/14 (1976–1977), pp. 15–72.

———. "Woman and Mathematics, Past and Present." *Philosophia Mathematica,* 13/14 (1976–1977), pp. 5–14.

Festschrift zur Gedächtnisfeier für Karl Weierstrass 1815–1965. Ed. Heinrich Behnke and Klaus Kopfermann. Köln: Westdeutscher Verlag, 1966.

Field, Daniel. *The End of Serfdom: Nobility and Bureaucracy in Russia 1855–1861*. Cambridge: Harvard University Press, 1976.

Figner, Vera. *Studencheskie gody, 1872–1876*. Moscow: "Golos truda", 1924.

Filippov, E. "S. V. Kovalevskaia." *Detskii otdykh*, No. 1 (1893), pp. 78–103.

"The First New Woman." *Post* (Pittsburgh), 27 May 1895.

Fleishits, Ek. "Pervyi vserossiiskii s"ezd po obrazovaniiu zhenshchin." *Vestnik Evropy*, No. 2 (1913), pp. 361–65.

Francais, Maurice. "Une Femme de Science. Chez Madame Kowaleski [sic]." *La Voltaire*, 1 July 1886.

Franklin, Fabian. " 'Masculine Heads' and 'Feminine Hearts'. Apropos of Sonya Kovalevsky." *Century Magazine*, 51 (1895–1896), pp. 317–18.

Freeze, Gregory L. *The Russian Levites: Parish Clergy in the Eighteenth Century*. Cambridge: Harvard University Press, 1977.

Frost, Laura. "Sonja Kowalewski. Ein Lebens und Charakterbild." *Deutsche Welt*, 12 and 19 February 1899.

G., E. [Barteneva?]. "Zhenskii kongress." *Novosti i birzhevaia gazeta*, 16 July 1889.

G., G. "S. V. Kovalevskaia. (Nekrolog)." *Russkie vedomosti*, 2 February 1916.

G., Ia. G. "A. N. Strannoliubskii. (Nekrolog)," *Russkaia shkola*, No. 5–6 (1903), pp. 89–90.

Gamble, Eliza Burt. *The Sexes in Science and History*. 1894; rpt. New York: G. P. Putnam's Sons, 1916.

"Le génie et le talent chez les femmes." *L'Italie*, 6 August 1893.

George, W. L. *The Intelligence of Woman*. Boston: Little, Brown, and Company, 1920.

Geronimus, J. L. *Sofja Wassiljevna Kowalewskaja—Mathematische Berechnung der Kreisbewegung*. Berlin: VEB Verlag Technik, 1954.

Gnedenko, B. V. and I. B. Pogrebysskii. *Mikhail Vasilevich Ostrogradskii, 1801–1862*. Moscow: AN SSSR, 1963.

Goldberg, Rochelle Lois. *The Russian Women's Movement: 1859–1917*. Diss. University of Rochester, 1976.

Goldenveizer, A. B. *Vblizi Tolstogo*. n. p.: Gos. Izd. Khudo. Lit., 1959.

Golitsyn, N. N. *Bibliograficheskii slovar' russkikh pisatel'nits*. St. Petersburg: Tip. V. S. Balasheva, 1889.

Golubev, V. V. "Raboty S. V. Kovalevskoi o dvizhenii tverdogo tela vokrug nepodvizhnoi tochki." *Prikladnaia matematika i tekhnika*, 14, No. 3 (1950), pp. 236–44.

Graham, Loren R. *Science and Philosophy in the Soviet Union*. New York: Vintage Books, 1974.

Gratsianskaia, L. N. "Elizaveta Fedorovna Litvinova." *Matematika v shkole*, No. 4 (1953), pp. 64–7.

Gumilevskii, Lev. *Chaplygin*. Moscow: Molodaia Gvardiia, 1969.

Hale, Sarah Josepha. *Woman's Record; Sketches of all Distinguished Women from the Creation to A.D. 1854*. New York: Harper & Bros., 1855.

Halpérine-Kaminsky, E. "Madame Lydia Yavorskaïa. Princesse Bariatinsky." *Le Théatre*, No. 9 (1902), pp. 19–20.

Hamilton, Louise. *Ellen Key. En Livsbild.* Stockholm: Wahlstrom & Widstrand, 1917.

Hansson, Laura Marholm. *Six Modern Women: Psychological Sketches.* Trans. Hermione Ramsden. Boston: Roberts Bros., 1896.

Hapgood, Isabel F. "Notable Women: Sonya Kovalevsky." *Century Magazine,* 50 (1895), pp. 536–39.

Holt, Alix, ed. *Selected Writings of Alexandra Kollontai.* New York: W. W. Norton & Company, 1977.

Iacobacci, Rora F. "Women of Mathematics." *Arithmetic Teacher,* 17 (1970), pp. 316–24.

Iakovkina, N. I. *Russkaia kul'tura vtoroi poloviny XIX veka.* Leningrad: Izd. Leningradskogo Universiteta, 1968.

Iankovskaia, Zh. *Sёstry.* Leningrad: Gos. Izd. Detskoi Literatury, 1962.

Ireland, Norma Olin. *Index to the Women of the World from Ancient to Modern Times: Biographies and Portraits.* Westwood, Mass.: F. W. Faxon, 1970.

Istoriia Otechestvennoi Matematiki. Vol. II. Kiev: Naukova Dumka, 1967.

Itenberg, B. S. *Rossiia i Parizhskaia kommuna.* Moscow: Nauka, 1971.

Iushkevich, A. "Vydaiushchiisia matematik Sofia Kovalevskaia." *Novyi mir,* No. 4 (1949), pp. 304–05.

Ivanich, Anna. "L'Ame Féminine Russe." *La Vie Féminine,* 18 November 1917.

Ivanovskii, I. *Maksim Maksimovich Kovalevskii. Biograficheskii ocherk.* Petrograd: Tip. B. M. Volfa, 1916.

"Iz dnevnika E. A. Shtakenshneider." *Golos minuvshego,* No. 11 (1915), pp. 161–203; No. 4 (1916), pp. 44–76.

"Izvlechenie iz protokolov zasedanii matematicheskogo obshchestva." *Matematicheskii sbornik,* 16 (1891–1892), pp. 827–48.

Joteyko, J. "A Propos des femmes mathematiciennes." *Revue Scientifique,* 9 January 1904, pp. 11–14.

"K biografii S. V. Kovalevskoi." *Novoe vremia,* 7 February 1891.

"K biografii Very Nikolaevny Figner." *Byloe,* No. 5 (1906), pp. 3–13.

"K kharakteristike S. V. Kovalevskoi." *Russkie vedomosti,* 5 February 1891.

K—o, V. "Sofia Kovalevskaia. Nekrolog." *Smolenskii vestnik,* 1 February 1891.

Kagan, Ven. "S. Kovalevskaia. Vospominaniia detstva i avtobiograficheskie ocherki." Rev. *Sovetskaia kniga,* No. 8–9 (1946), pp. 88–90.

Kannak, E. "S. Kovalevksaia i M. Kovalevskii." *Novyi zhurnal,* No. 39 (December 1954), pp. 194–211.

Kazakov, A. P. *Teoriia progressa v russkoi sotsiologii kontsa XIX veka. (P. L. Lavrov, N. K. Mikhailovskii, M. M. Kovalevskii).* Leningrad: Izd. Leningradskogo Universiteta, 1969.

Kennan, George. *Siberia and the Exile System.* 1891; rpt. Chicago: The University of Chicago Press, 1958.

Kerbedz, E. de. "Sophie de Kowalevski." *Rendiconti del Circulo Matematico di Palermo,* 5, 12 April 1891, pp. 121–28.

Key, Ellen. *Drei frauenschicksale.* 2nd ed. Berlin: S. Fischer Verlag, 1908.

———. *The Renaissance of Motherhood.* Trans. Anna E. B. Fries. New York: G. P. Putnam's Sons, 1914.

―――――. *The Woman Movement*. Trans. Mamah Bouton Borthwick. New York: G. P. Putnam's Sons, 1912.

Khvostov, V. "Voina i zhenskii vopros." *Russkie vedomosti*, 9 March 1916.

Kimberling, Clark H. "Emmy Noether." *American Mathematical Monthly*, 79, No. 2 (1972), pp. 136–49.

Klein, Felix. *Vorlesungen über die Entwicklung der Mathematik im 19. Jahrhundert.* Berlin: Springer, 1926. Grundlehren No. 24.

Kline, Morris. *Mathematical Thought from Ancient to Modern Times.* New York: Oxford University Press, 1972.

Kniazev, G. A. "O rukopisnom nasledstve S. V. Kovalevskoi." *Vestnik Akademii Nauk SSSR*, No. 6 (1944), pp. 96–7.

Knizhnik-Vetrov, I. S. *A. V. Korvin-Krukovskaia (Zhaklar). Drug F. M. Dostoevskogo i deiatel'nitsa parizhskoi kommuny.* Moscow: Izd. Vsesoiuznogo Obshchestva, 1931.

―――――. *Russkie deiatel'nitsy pervogo internatsionala i parizhskoi kommuny. E. L. Dmitrieva, A. V. Zhaklar, E. G. Barteneva.* Moscow: Nauka, 1964.

Koblitz, Ann Hibner. "Science, Women, and Revolution in Russia." *Science for the People*, 14, No. 4 (1982), pp. 14–8; 34–7.

Kolbin, Georgii. "Osnovatel' novoi paleontologii." *Ogonëk*, No. 33 (1949), pp. 21–2.

Koltonovskaia, E. A. "Bol'shaia dusha." *Rech'*, 29 January 1916.

―――――. *Zhenskie siluety.* St. Petersburg: Prosveshchenie, 1912.

Korvin-Krukovskaia, A. V. "Mikhail." *Epokha*, No. 9 (1864), pp. 1–58. [Under the pseudonym Iu. O—v]

―――――. "Son'." *Epokha*, No. 8 (1864), pp. 1–24.

Korvin-Krukovskii, F. V. "Sofia Vasilevna Korvin-Krukovskaia." *Russkaia starina*, 71, No. 9 (1891), pp. 623–36.

Kotliarevskii, Nestor. Rev. of "Nigilistka." *Strana*, 3 September 1906.

Kovalevskaia-Chistovich, V. A. "Aleksandr Onufrievich Kovalevskii. (Vospominaniia docheri)." *Priroda*, No. 7–8 (1926), pp. 5–20.

Kovalevskii, A. O. *Pis'ma A. O. Kovalevskogo k I. I. Mechnikovu, 1866–1900.* Ed. Iu. I. Polianskii et al. Moscow: AN SSSR, 1955.

―――――, and V. O. Kovalevskii. "V. O. i A. O. Kovalevskie [letters]." *Nauchnoe nasledstvo.* Moscow: AN SSSR, 1948. Vol. I, pp. 185–423. (NN)

Kovalevskii, M. M. "Moe nauchnoe i literaturnoe skital'chestvo." *Russkaia mysl'*, No. 1 (1895), pp. 61–80.

―――――. "Moskovskii universitet v kontse 70-kh i nachale 80-kh godov proshlogo veka (lichnyia vospominaniia)." *Vestnik Evropy*, No. 5 (1910), pp. 171–221.

―――――. *Ocherk proizkhozhdeniia i razvitiia sem'i i sobstvennosti.* St. Petersburg: Tip. Iu. N. Erlikh, 1896.

―――――. "Vysshee zhenskoe obrazovanie." *Vestnik Evropy*, No. 6 (1911), pp. 416–25.

―――――. "Za rubezhom: Iz perepiski russkikh deiatelei za granitsei: Gertsena, Lavrova i Turgeneva." *Vestnik Evropy*, No. 3 (1914), pp. 210–30.

Kovalevskii, Vladimir. *Zametka o moem magisterskom ekzamene.* Kiev: Universitetskaia Tip., 1874.

Kramer, Edna E. *The Main Stream of Mathematics.* New York: Oxford University Press, 1951.

Kronecker, L. "Sophie von Kowalevsky." *Crelle's Journal,* 108 (1891), p. 88.

Kropotkin, Peter. *Memoirs of a Revolutionist.* 1900; rpt. New York: Grove Press, 1968.

Krylov, A. N. *Vospominaniia i Ocherki.* Moscow: Voennoe Izd., 1949.

Kuskova, Ek. "Sofia Nikolaevna [sic] Kovalevskaia i ee vremia." *Golos minuvshego,* No. 2, pp. 215–26.

––––––. "Sofia Vasilevna Kovalevskaia." *Russkie vedomosti,* 29 January 1916.

Kuznetsov, B. G. "Iz proshlogo russkoi nauki." *Novyi mir,* No. 8 (1938), pp. 175–96.

L., E. "Sophie Kowalewska." *Bulletin de l'Union Universelle des femmes,* No. 14, 15 February 1891, pp. 1–4.

Lange, Helene. *Higher Education of Women in Europe.* Trans. L. R. Klemm. New York: D. Appleton and Company, 1897.

––––––. "Sophie von Kowalevsky." *Die Nation,* No. 22, 28 February 1891, pp. 340–42.

Laurent, Eugène. "La Genèse de la Pitié." *Revue Internationale de Sociologie,* No. 10 (1897), pp. 733–38.

Lavrov, P. L. "Pis'ma P. L. Lavrova k E. A. Shtakenshneider iz Parizha v 1870–1873g." *Golos minuvshego,* No. 7–8 (1916), pp. 107–40; No. 9, pp. 114–38.

––––––. *Russkaia razvitaia zhenshchina. V pamiati S. V. Kovalevskoi.* Geneva: Volnaia Russkaia Tipografiia, 1891.

Lazarev, P. P. *Ocherki istorii russkoi nauki.* Moscow: Izd. AN SSSR, 1950.

Ledkovsky, Marina. "Avdotya Panaeva: Her Salon and Her Life." *Russian Literature Triquarterly,* No. 9 (Spring 1974), pp. 423–32.

Lemke, Mikhail, ed. *M. M. Stasiulevich i ego sovremenniki v ikh perepiske.* St. Petersburg: Tip. M. M. Stasiulevicha, 1912. Vol. II.

––––––. *Ocherki osvoboditel'nogo dvizheniia "shestidesiatykh godov."* St. Petersburg: Izd. O. N. Popovoi, 1908.

Leningradki. Vospominaniia. Ocherki. Dokumenty. Leningrad: Lenizdat, 1968.

Lestvitsyn, V. "Ot"ezd A. M. Evreinovoi." *Iaroslavskie gubernskie vedomosti,* 18 September 1880.

Liapunov, A. M. *Sobranie sochinenii.* Moscow: AN SSSR, 1954.

Likhacheva, E. O. *Materialy dlia istorii zhenskogo obrazovaniia v Rossii. (1086–1901).* 6 vol. St. Petersburg: Tip. M. M. Stasiulevicha, 1893–1901.

Linder, Gurli, "Sonja Kovalevsky." *Ord och Bild,* No. 23 (1930), pp. 347–64.

Litvinova, E. F. "Iz vremen moego studenchestva. (Znakomstvo s S. V. Kovalevskoi)." *Zhenskoe delo,* No. 4 (1899), pp. 34–63. [under the pseudonym E. El']

––––––. *N. I. Lobachevskii, ego zhizn' i uchenaia deiatel'nost'.* St. Petersburg: "Obshchestvennaia Pol'za", 1894.

––––––. *S. V. Kovalevskaia: ee zhizn' i nauchnaia deiatel'nost'.* St. Petersburg: "Obshchestvennaia Pol'za", 1893.

Liubatovich-Dzhabadari, Olga. "Dalekoe i nedavnee." *Byloe,* No. 5 (1906), pp. 209–45; No. 6, pp. 108–54.

Liudi Russkoi Nauki. 2 vol. Moscow: Gos. Izd. Tekh.-Teor. Lit., 1948.
Lockemann, Georg. *Robert Wilhelm Bunsen.* Stuttgart: Wissenschaft Verlags-gesellschaft MBH, 1949.
Lombroso, César. "Le Génie et le Talent chez les Femmes." *Revue des Revues,* August 1893, pp. 561–67.
Loria, Gino. "Donne mathematiche." *Scritti, Conferenze, Discorsi sulla Storia delle Matematiche.* Padua: n. p., 1937, pp. 447–66.
———. "Encore les femmes mathématiciennes." *Revue Scientifique,* 5, No. 1 (1904), pp. 338–40.
———. "Les femmes mathématiciennes." *Revue Scientifique,* 20, No. 13 (1903), pp. 387–92.
Luchins, Edith H. and Abraham S. Luchins. "Female Mathematicians: A Contemporary Appraisal." in *Women and the Mathematical Mystique.* Ed. Lynn H. Fox, et al. Baltimore: The Johns Hopkins University Press, 1980.
Lukomskaia, A. M. *Aleksandr Mikhailovich Liapunov. Bibliografiia.* Moscow: AN SSSR, 1953.
M. M. Kovalevskii 1851–1916. Sbornik statei. Petrograd: Tip. A. F. Marks, 1918.
"A Magazine Causerie." *The Illustrated London News,* 18 May 1895.
"Malen'kii fel'eton o russkoi zhenshchine." *Volyn',* 13 September 1897.
Malevich, Iosif Ignatievich. "Sofia Vasilevna Kovalevskaia, doktor filosofii i professor vysshei matematiki, v 'Vospominiiakh pervogo, po vremeni, ee uchitelia I. I. Malevicha.' " *Russkaia starina,* No. 12 (1890), pp. 615–54.
Matematika XIX veka. Ed. A. N. Kolmogorov and A. P. Iushkevich. Moscow: Nauka, 1981.
Matveev, N. *Sofia Kovalevskaia. Printsessa nauki. Povest' o zhizni.* Moscow: Molodaia Gvardiia, 1979.
Mechnikov, I. I. "L'individu chez les animaux et dans l'humanité." *La Revue,* 15 April 1904, pp. 389–98.
———. *Stranitsy vospominanii. Sbornik avtobiograficheskikh statei.* Moscow: AN SSSR, 1946.
Mechnikova, O. N. "Druzhba mezhdu A. O. Kovalevskim i Il. Il. Mechnikovym." *Priroda,* No. 7–8 (1926), pp. 31–8.
Meijer, J. M. *Knowledge and Revolution: The Russian Colony in Zürich (1870–1873).* Assen: Van Gorcum, 1955.
Mendelson, Marie. "Briefe von Sophie Kowalewska." *Neue Deutsche Rundschau,* No. 6 (1897), pp. 589–614.
———. "Vospominaniia o Sofe Kovalevskoi." Trans. L. Krukovskaia. *Sovremennyi mir,* No. 2 (1912), 134–76.
Metelkin, A. I. "S. V. Kovalevskaia." Rev. of *Vospominaniia detstva. Priroda.* No. 6 (1946), pp. 95–6.
Meyer, Alfred G. "Marxism and the Women's Movement." in *Women in Russia.* Ed. Dorothy Atkinson et al. Stanford: Stanford University Press, 1977, pp. 85–112.
Mikhailevich, Vsevolod. "Pamiati S. V. Kovalevskoi." *Volzhskii vestnik,* 12 February 1899.
Mirovich, N. "Sofia Kovalevskaia." *Zhenskoe delo,* 10 June 1910.
Mittag-Leffler, G. "Une page de le vie de Weierstrass." *Compte rendu du*

deuxième congrès international des mathématiciens. Paris: Gauthier-Villars, 1902, pp. 131–53.

———. "Sophie Kovalevsky, notice biographique." *Acta Mathematica,* 16 (1893), pp. 385–92.

———. "Weierstrass et Sonja Kowalewsky." *Acta Mathematica,* 39 (1923), pp. 133–98.

Molchanov, G. F. "Sofia Vasilevna Kovalevskaia." *Nauka i zhizn',* No. 1 (1950), p. 42.

Moseley, Maboth, *Irascible Genius: The Life of Charles Babbage.* Chicago: Henry Regnery Company, 1964.

Moser, Charles A. *Antinihilism in the Russian Novel of the 1860s.* The Hague: Mouton, 1964.

Mozans, H. J. *Woman in Science.* 1913; rpt. Cambridge: The MIT Press, 1974.

Musabekov, Iu. S. *Iulia Vsevolodovna Lermontova 1846–1919.* Moscow: Nauka, 1967.

Naimark, Norman M. *The History of the "Proletariat": The Emergence of Marxism in the Kingdom of Poland, 1870–1887.* No. LIV of East European Monographs. New York: East European Quarterly, 1979.

Natkowski, Waclaw. "Das Tagebuch der Kowalewska." *Wiener Rundschau,* 15 (1901), pp. 32–9.

Nekrasova, E. S. "Zhenskie vrachebnye kursy v Peterburge. Iz vospominanii i perepiski pervykh studentok." *Vestnik Evropy,* No. 6 (1882), pp. 807–45.

"Nekrolog. Sofia Vasilevna Kovalevskaia." *Kharkovskie gubernskie vedomosti,* 1 February 1891.

Nikitenko, S. "Zhenshchiny-professora Bolonskogo universiteta." *Russkaia mysl',* No. 10 (1883), pp. 253–91; No. 11, pp. 85–127.

Novarese, E. "Sofia Kowalevski. Nekrolog." *Rivista di Matematica,* No. 2–3 (1891), pp. 21–2.

Obolenskii, L. E. "Kovalevskaia i Bashkirtseva. (Stranichka iz zhenskoi zhizni vysshego tipa." *Knizhki nedeli,* No. 2 (1893), pp. 148–67.

Obolianinova, S. A. *Istoricheskii Put' Russkoi Zhenshchiny.* New York: Tip. Gazety "Rossiia", 1968.

Odna iz mnogikh. Iz zapisok nigilistki. St. Petersburg: Tip. Ed. Mettsiga, 1881.

Oleinik, O. A. "Teorema S. V. Kovalevskoi i ee rol' v sovremennoi teorii uravnenii s chastnymi proizvodnymi." *Matematika v shkole,* No. 5 (1975), pp. 5–9.

Osen, Lynn M. *Women in Mathematics.* Cambridge, Mass.: The MIT Press, 1974.

Ozhigova, E. P. "P. L. Chebyshev i Peterburgskaia Akademiia nauk." *Voprosy istorii estestvoznaniia i tekhniki,* 4(49) (1974), pp. 15–21.

———. *Sharl' Ermit, 1822–1901.* Leningrad: Nauka, 1982.

P., T. "A Book of the Week. Sophie Kovalesky [sic]." *The Weekly Sun,* 23 and 30 September 1894.

"Pamiati S. V. Kovalevskoi." *Birzhevye vedomosti,* 30 January 1916.

Pamiati S. V. Kovalevskoi. Sbornik statei. Ed. P. Ia. Polubarinova-Kochina. Moscow: AN SSSR, 1951.

"Pamiatnik S. V. Kovalevskoi." *Orlovskii vestnik,* 20 September 1896.

Panaeva, A. Ia. *Vospominaniia*. Moscow: Gos. Izd. Khudo. Lit., 1956.

Panteleev, L. F. *Iz vospominanii proshlogo*. 1905; rpt. Moscow: Academia, 1934.

———. "Iz zhizni S. V. Kovalevskoi." *Birzhevye vedomosti*, 29 January 1916.

———. "Pamiati S. V. Kovalevskoi." *Rech'*, 29 January 1916.

———. *Vospominaniia*. n. p.: Gos. Izd. Khudo. Lit., 1958.

Payne, Robert. *Dostoevsky: A Human Portrait*. New York: Alfred A. Knopf, 1967.

Pelageia Iakovlevna Kochina. Moscow: Nauka, 1977.

Perl, Teri. *Math Equals. Biographies of Women Mathematicians + Related Activities*. Menlo Park, Ca.: Addison-Wesley Publishing Company, 1978.

Peskovskii, M. "Universitetskaia nauka dlia russkikh zhenshchin." *Russkaia mysl'*, No. 11 (1886), pp. 42–64; No. 12, pp. 1–30.

Picard, Emile. *O pazvitii za poslednie sto let nekotorykh osnovnykh teorii matematicheskogo analiza*. Trans. S. N. Bernshtein. Kharkov: Tip. M. Zilberberg, 1912.

Pisareva, E. *Pamiati Anny Pavlovny Filosofovoi*. St. Petersburg: Izdanie Rossiiskogo Teosoficheskogo Obshchestva, 1912.

Pis'ma A. P. Chekhova. Vol. III. Moscow: Knigoizdatel'stvo pisatelei, 1915.

"Pis'ma S. V. Kovalevskoi 1868 g." *Golos minuvshego*, No. 2 (1916), pp. 226–40; No. 3, pp. 213–31; No. 4, pp. 77–94.

Pogozhev, A. V. *Dvadtsatipiatiletie estestvenno-nauchnykh s"ezdov v Rossii, 1861–1886*. Moscow: Tip. V. M. Frish, 1887.

Poincaré, Henri. *The Foundations of Science*. Trans. George Bruce Halsted, 1913 rpt. Lancaster, Pa.: The Science Press, 1946.

———. *Mathematics and Science: Last Essays*. Trans. John W. Bolduc, 1913 rpt. New York: Dover, 1963.

Pokrovskii, K. D. "Zhenshchiny-astronomy i ikh raboty." *Istoricheskii vestnik*, 68 (1897), pp. 761–79.

Polnoe sobranie sochinenii P. L. Chebysheva. Moscow: AN SSSR, 1951. Vol. 5.

Polubarinova-Kochina, P. Ia. "Dzh. Dzh. Silvestr i S. V. Kovalevskoi." *Voprosy istorii estestvoznaniia i tekhniki*, 5 (1957), pp. 156–62.

———. "Iz perepiski S. V. Kovalevskoi." *Uspekhi matematicheskikh nauk*, 7, No. 4(50) (1952), pp. 103–25.

———. "K biografii S. V. Kovalevskoi." *Istoriko-matematicheskie issledovaniia*, 7 (1954), pp. 666–712.

———. "Karl Teodor Vilgelm Veierstrass (k 150-letiiu so dnia rozhdeniia)." *Uspekhi matematicheskikh nauk*, 21, No. 3(129) (1966), pp. 213–24.

———. "Nauchnye raboty S. V. Kovalevskoi. K stoletiiu so dnia rozhdeniia." *Prikladnaia matematika i mekhanika*, 14, No. 3, (1950), pp. 229–35.

———. *Nikolai Evgrafovich Kochin 1901–1944*. Moscow: Nauka, 1979.

———. *Pis'ma S. V. Kovalevskoi ot inostrannykh korrespondentov*. Preprint No. 121. Moscow: Institut problem mekhaniki AN SSSR, 1979.

———. *Sofia Vasilevna Kovalevskaia 1850–1891*. Moscow: Nauka, 1981.

———. *Sofia Vasilevna Kovalevskaia. Ee zhizn' i deiatel'nost'*. Moscow: Gos. Izd. Tekh.-Teor. Lit., 1955.

———. "Sofia Vasilevna Kovalevskaia. (Ocherk nauchnoi deiatel'nosti)." in Kovalevskaia, *Nauchnye raboty*. Moscow: AN SSSR, 1948, pp. 313–42.

———. "V zashchitu Sofi Kovalevskoi." *Pravda*, 8 July 1948.
———. *Zhizn' i deiatel'nost' S. V. Kovalevskoi*. Moscow: AN SSSR, 1950.
Pomper, Philip. *Peter Lavrov and the Russian Revolutionary Movement*. Chicago: The University of Chicago Press, 1972.
———. *Sergei Nechaev*. New Brunswick: Rutgers University Press, 1979.
Ponomarev, S. I. *Nashi pisatelnitsy*. St. Petersburg: Tip. Imp. Aka. Nauk, 1891.
Povorinskaia, V. "Sofia Vasilevna Kovalevskaia. Biograficheskii Ocherk." *Rodnik*, No. 6 (1900), pp. 607–16.
Prozor, M. le comte M. "Femmes du Nord: Sophie Kowalewsky, Anne Charlotte Leffler, Duchesse de Caianello." *La Vie Contemporaine*, 3, August 1893, pp. 297–314.
Prudnikov, V. E. *Pafnutii Lvovich Chebyshev 1821–1894*. Leningrad: Nauka, 1976.
———. "S. V. Kovalevskaia i P. L. Chebyshev (K 100-letiiu so dnia rozhdeniia S. V. Kovalevskoi." *Priroda*, No. 4 (1950), pp. 72–6.
Pypin, A. N. "Literaturnoe obozrenie: Literaturnye sochineniia S. V. Kovalevskoi." *Vestnik Evropy*, No. 1 (1893), pp. 889–94.
Radovskii, M. I. Rev. of "S. Ia. Shtraikh. Sem'ia Kovalevskikh." *Priroda*, No. 4 (1950), pp. 93–6.
———. Rev. of "S. V. Kovalevskaia. Vospominaniia detstva i avtobiograficheskie ocherki." *Nauka i zhizn'*, No. 2–3 (1946), pp. 44–6.
Raeff, Marc. *Origins of the Russian Intelligentsia: The Eighteenth-Century Nobility*. New York: Harcourt, Brace & World, 1966.
Rappaport, Karen D. "S. Kovalevsky: A Mathematical Lesson." *American Mathematical Monthly*, 88, No. 8 (1981), pp. 564–74.
Rebière, A. *Les Femmes dans la Science*. Paris: Nony et Cie, 1894.
———. *Mathématiques et Mathematiciens*. Paris: Nony, 1893.
———. *Les Savants Modernes*. Paris: Nony, 1899.
Reich, Emil. *Woman Through the Ages*. London: Methuen & Co., 1908. Vol. II.
Reid, Robert. *Marie Curie*. New York: E. P. Dutton, 1974.
"A Remarkable Russian Woman. Reminiscences of Mme. Sophie Kovalevsky, late Professor of Mathematics at the Stockholm University. By a lady friend of hers." *Free Russia* (New York), 8 April 1891.
"Reviews." Rev. of Kovalevskaia's *A Russian Childhood*. Trans. Beatrice Stillman. *Mathematics Magazine*, 52, No. 3 (1979), p. 186.
Reznik, Semen. *Mechnikov*. Moscow: Molodaia Gvardiia, 1973.
———. *Vladimir Kovalevskii (Tragediia nigilista)*. Moscow: Molodaia Gvardiia, 1978.
Rod Kovalevskikh za trista let 1651–1951. Paris: Privately printed, 1951.
Rogger, Hans. "Russia." in *The European Right: A Historical Profile*. Ed. Hans Rogger and Eugen Weber. Berkeley: University of California Press, 1966, pp. 443–500.
Rostov, N. "Russkie zhenshchiny v Parizhskoi kommuny." *Ogonëk*, No. 10–11 (1946), pp. 16–7.
Rowbotham, Sheila. *Women, Resistance and Revolution*. London: Allen Lane—The Penguin Press, 1972.
Russian, Ts. K. "S. V. Kovalevskaia (v dvadtsatipiatiletniuiu godovshchinu

smerti." *Soobshcheniia Kharkovskogo matematicheskogo obshchestva*, 15, No. 4 (1917), pp. 161–72.

S., A. "Pamiati Sofi Vasilevny Kovalevskoi." *Zhenskoe obrazovanie*, No. 2 (1891), pp. 213–17.

Sarcey, Francisque. "Les Livres." *La Revue Illustrée*, 15 April 1895, pp. 317, 319.

Satina, Sophie. *Education of Women in Pre-Revolutionary Russia*. Trans. Alexandra F. Poustchine. New York: n.p., 1966.

Schirmacher, Dr. Kaethe. *The Modern Women's Rights Movement: A Historical Survey*. Trans. C. C. Eckhardt. New York: The Macmillan Company, 1912.

Schlözer, Leopold von. *Dorothea von Schlözer, der Philosophie Doctor*. Berlin: Deutsche Verlags, 1925.

Sechenov, I. M. *Avtobiograficheskie zapiski*. Moscow: AN SSSR, 1952.

Semenov, D. "Iz perezhitogo. V Mariinskoi zhenskoi gimnazii." *Russkaia shkola*, No. 3 (1892), pp. 24–38.

Semevskii, M. I. "Putevye ocherki, zametki i nabroski. Poezdka po Rossii v 1890 g." *Russkaia starina*, 68, No. 12 (1890), pp. 713–45.

Semevskii, V. I. "Avtobiograficheskie nabroski V. I. Semevskogo." *Golos minuvshego*, No. 9–10 (1917), pp. 7–49.

Shashkov, S. S. *Sobranie sochinenii*. St. Petersburg: Tip. I. N. Sorokhodova, 1898. Vol. I.

Shelgunov, N. V., L. P. Shelgunova, and M. L. Mikhailov. *Vospominaniia*. Moscow: Izd. Khudo. Lit., 1967. 2 vol.

Sheremetevskaia, A. N. "Stranitsa iz istorii vysshego zhenskogo obrazovaniia." *Istoricheskii vestnik*, 65 (1896), pp. 171–84.

Shtakenshneider, E. A. *Dnevnik i zapiski (1854–1886)*. Moscow: Academia, 1934.

Shtraikh, S. Ia. "A. I. Gertsen i V. O. Kovalevskii." *Istoricheskie Zapiski*, 54 (1955), pp. 448–63.

———. "Dostoevskii i sestry Korvin-Krukovskie." *Krasnaia nov'*, No. 7 (1931), pp. 144–50.

———. "Genial'nyi russkii paleontolog Vladimir Kovalevskii." *Nauka i zhizn'*, No. 10 (1942), pp. 39–42.

———. "N. I. Pirogov o liubvi, o prizvanii zhenshchiny—materi i pr." *Golos minuvshego*, No. 5 (1915), pp. 189–206; No. 6, pp. 191–205.

———. "Russkaia nigilistka v Parizhskoi kommune." *Molodaia gvardiia*, No. 5–6 (1931), pp. 114–20.

———. *S. Kovalevskaia*. Moscow: Zhurn.-Gaz. Ob"edinenie, 1935.

———. *Sem'ia Kovalevskikh*. Moscow: Sovetskii Pisatel', 1948.

———. *Sestry Korvin-Krukovskie*. Moscow: Mir, 1933.

Sigma [S. N. Syromiatnikov]. "Drama znamenitoi zhenshchiny." *Novoe vremia*, 21 April 1895.

Skabichevskii, A. M. *Istoriia noveishei russkoi literatury: 1848–1892 gg*. 2nd ed. St. Petersburg: Izd. F. Pavlenkova, 1893.

———. *Sochineniia*. St. Petersburg: Tip. Iu. N. Erlikh, 1903. 2 vol.

Skriba, P. "Iz biografii S. V. Kovalevskoi." *Russkaia zhizn'*, 5 February and 11 March 1893.

Sletova, Vera. "Vydaiushchiisia matematik." *Sovetskaia zhenshchina*, No. 2 (1946), pp. 55–6.

Smirnov, German. *Mendeleev*. Moscow: Molodaia Gvardiia, 1974.

Smirnov, V. I. "Russkaia matematika XIX i XX vekov." *Priroda*, No. 3 (1945), pp. 17–23.

Smith, David Eugene. *History of Mathematics*. Boston: Ginn and Company, 1923. Vol. I.

————, and Jekuthiel Ginsburg. *A History of Mathematics in America Before 1900*. Carus Mathematical Monograph No. 5. Chicago: The Mathematical Association of America, 1934.

Sobranie sochinenii Georga Brandesa. Trans. M. V. Luchitskaia. Kiev: Izd. B. K. Fuksa, 1902.

"Some Very Modern Women." *The Saturday Review*, 29 February 1896.

Somerville, Martha. *Personal Recollections from Early Life to Old Age of Mary Somerville*. Boston: Roberts Bros., 1876.

Sonin, N. Ia. "Zametka po povodu pis'ma P. L. Chebysheva k S. V. Kovalevskoi." *Izvestiia Imperatorskoi Akademii Nauk*, 2, No. 1 (1895) pp. 15–26.

Stanton, Theodore, ed. *The Woman Question in Europe*. New York: G. P. Putnam's Sons, 1884.

Stasov, Vladimir. *Nadezhda Vasilevna Stasova: Vospominaniia i ocherki*. St. Petersburg: Tip. M. Merkusheva, 1899.

Stepniak-Kravchinsky, S. *King Stork and King Log. A Study of Modern Russia*. London: Downey & Co., 1895. 2 vol.

Stillman, Beatrice. "Sofya Kovalevskaya: Growing Up in the Sixties." *Russian Literature Triquarterly*, No. 9 (Spring 1974), pp. 276–302.

Stites, Richard. "M. L. Mikhailov and the Emergence of the Woman Question in Russia." *Canadian Slavic Studies*, 3, No. 2 (Summer 1969), pp. 178–99.

————. "Women and the Russian Intelligentsia: Three Perspectives." in *Women in Russia*. Ed. Dorothy Atkinson, et al. Stanford: Stanford University Press, 1977, pp. 39–62.

————. *The Women's Liberation Movement in Russia: Feminism, Nihilism, and Bolshevism 1860–1930*. Princeton: Princeton University Press, 1978.

Stoiunin, V. Ia. "Obrazovanie russkoi zhenshchiny (Po povodu dvadtsatipiatiletiia russkikh zhenskikh gimnazii)." *Istoricheskii vestnik*, 12 (1883), pp. 125–53.

Stoletov, A. G. "Russkaia zhenshchina-matematik S. V. Kovalevskaia." *Nauka i zhizn'*, No. 23 (1891), pp. 353–56.

————, N. E. Zhukovskii, and P. A. Nekrasov. "S. V. Kovalevskaia." *Matematicheskii sbornik*, 16 (1891–1892), pp. 1–38.

"Stranitsa v Al'bome 'Russkoi Stariny'." *Russkaia starina*, 71, No. 9 (1891), pp. 640–41.

"Strannaia smert'. Samoubiistvo V. O. Kovalevskogo." *Novoe vremia*, 21 April 1883.

Strannoliubskii, A. "Zhenskoe obrazovanie v Rossii." *Obrazovanie*, No. 10 (1894), pp. 320–44.

Suslova, A. P. *Gody blizosti s Dostoevskim.* Moscow: M. & S. Sabashnikovykh, 1928.

Suslova, N. P. "Iz nedavnego proshlogo." *Vestnik Evropy*, No. 3 (1900) pp. 624–73.

Svatikov, S. G. "Opal'naia professura 80 gg." *Golos minuvshego*, No. 2 (1917), pp. 5–78.

————. "Russkie studenty v Geidelberge." *Novyi zhurnal dlia vsekh*, No. 12 (1912), pp. 69–82.

Sylvester, J. J. "Sonnet." *Nature*, 9 December 1886, p. 132.

Tarasov, D. A. "Professor S. V. Kovalevskaia (Po povodu desiatiletiia so dnia ee konchiny)." *Russkie vedomosti*, 30 January 1901.

Tee, G. J. "Sof'ya Vasil'yevna Kovalevskaya." *Mathematics Chronicle*, 5 (1977), pp. 113–39.

Tikhomandritskii, M. A. "Karl Weierstrass." *Soobshcheniia Kharkovskogo matematicheskogo obshchestva*, 6, No. 1–3 (1897), pp. 35–56.

Tikhomirov, Lev A. *Nachala i kontsy: "Liberaly" i Terroristy.* Moscow: Universitetskaia Tipografiia, 1890.

Timiriazev, K. A. *Nauka i demokratiia. Sbornik statei 1904–1919 gg.* 1920; rpt. Moscow: Izd. Sotsial'no-Ekonomicheskoi Lit., 1963.

Tolstoi, Lev, and V. V. Stasov. *Perepiska 1878–1906.* Leningrad: Priboi, 1929.

Tovrov, Jessica. "Mother-Child Relationships among the Russian Nobility." in *The Family in Imperial Russia.* Ed. David L. Ransel. Urbana: University of Illinois Press, 1978, pp. 15–43.

Tsiolkovskii, K. E. "Iz moei zhizni." *Izvestia* and *Pravda*, 20 September 1935.

Tuchkova-Ogareva, N. A. *Vospominaniia.* Moscow: Gos. Izd. Khudo. Lit., 1959.

Ulam, Adam B. *In the Name of the People. Prophets and Conspirators in Prerevolutionary Russia.* New York: The Viking Press, 1977.

"V". "Teatr' i musika." *Novoe vremia*, 20 April 1895.

V., Z. "Pamiatnik S. V. Kovalevskoi." *Novoe vremia*, 11 October 1896.

Valk, S. N. ed. *Sankt-Peterburgskie vysshie (Bestuzhevskie) kursy (1878–1918). Sbornik statei.* Leningrad: Izd. Leningradskogo Universiteta, 1973. 2nd ed.

Vartho. "S. V. Kovalevskaia (K 25-letiiu konchiny)." *Den'*, 29 January 1916.

Vasilev, A. V. "Pamiati Sofii Vasilevny Kovalevskoi." *Rech'*, 29 January 1916.

"Vecher v pamiat' S. V. Kovalevskoi." *Volzhskii vestnik*, 17 April 1891.

Venturi, Franco. *Roots of Revolution.* Trans. Francis Haskell. New York: Grosset & Dunlap, 1966.

Vil'e, Baron Mark de. *Zhenskie kluby i legiony amazonok.* Moscow: "Sovremennye problemy", 1912.

Vitkovskii, V. V. *Perezhitoe.* Leningrad: n.p., 1927.

Vodovozova, E. N. *Na zare zhizni.* Moscow: Izd. Khudo. Lit., 1964. 2 vol.

Volodin, A., and B. Itenberg. *Lavrov.* Moscow: Molodaia Gvardiia, 1981.

Volterra, Vito. "Sur les vibrations lumineuses dans les milieux biréfringents." *Acta Mathematica*, 16 (1892–1893), pp. 153–215.

Volynskii, A. "Literaturnye zametki." *Severnyi vestnik*, No. 10 (1890), pp. 153–67; No. 3 (1891), pp. 147–63.

Vorontsov-Veliaminov, B. A. *Ocherki o vselennoi*. Moscow: Izd. Tekh.-Teor. Lit., 1955.

Vorontsova, L. A. *Sofia Kovalevskaia*. Moscow: Molodaia Gvardiia, 1957.

"Vospominaniia Marii Konstantinovny Tsebrikovoi. Dvadtsatipiatiletie zhenskogo voprosa 1861–1886 gg." *Zvezda*, No. 6 (1935), pp. 190–208.

Vpered! Sbornik statei posviashchennykh pamiati Petra Lavrovicha Lavrova. Petrograd: Kolos, 1920.

Vucinich, Alexander. *Science in Russian Culture 1861–1917*. Stanford: Stanford University Press, 1970.

Vyrubov, G. N. "Revoliutsionnye vospominaniia. (Gertsen, Bakunin, Lavrov)," *Vestnik Evropy*, No. 1 (1913), pp. 53–79; No. 2, pp. 45–70.

———. "Voennye vospominaniia." *Vestnik Evropy*, No. 1 (1911) pp. 3–43; No. 2, pp. 92–125.

W., M. L. "Sophia Kovalevsky, The Russian New Woman. A Cinncinnati Writer Finds Proof in Her Sad Life That Woman's Sphere Is Home." *The Gazette*, 7 September 1895.

Webster, Arthur Gordon. Letter. *The Nation*, 1 February 1894, p. 83.

Weil, André. "Mittag-Leffler as I remember him." *Acta Mathematica*, 148 (1982), pp. 9–13.

Wenguerow, Zenaide. "La Femme Russe." *La Revue des Revues*, 15 September 1897, pp. 489–99.

Wentscher, M. "Weierstrass und Sonja v. Kowalewsky." *Jahresbericht der deutschen Matematiker-Vereinigung*, 18, No. 2 (1909), pp. 89–93.

Weyl, Hermann. "Emmy Noether." *Scripta Mathematica*, 8, No. 3 (July 1935), pp. 201–20.

Wlada, Julia. "Les Femmes Astronomes." *La Tribune de la Fronde*, 26 and 27 December 1898.

"Women as Mathematicians and Astronomers." *American Mathematical Monthly*, 25, No. 3 (1918), pp. 136–39.

"Women Eminent in Mathematics." *American Mathematical Monthly*, 44, No. 1 (1937), pp. 46–7.

Zabrezhnev, Iv. "U finnov i shvedov." *Novoe vremia*, 15 September 1900.

Zanta, Leontine. "Les Carrières Féminines." *La Vie Féminine*, 14 July 1914.

Zavadskii, K. M. *Razvitie evoliutsionnoi teorii posle Darvina. 1859–1920'e gody*. Leningrad: Nauka, 1973.

Zednik, Jella von. *Sophie Kowalewsky, ein weiblicher Professor*. Prague: Sammlung Gemeinnutziger Vorträge, 1898.

Zenkevich, I. G. *Sud'ba talanta. Ocherki o zhenshchinakh-matematikakh*. Briansk: Pedagogicheskoe Obshchestvo RSFSR, Brianskoe Obl. Otdelenie, 1968.

Zhukovskaia, Ekaterina. *Zapiski*. Leningrad: Izd. Pisatelei, 1930.

Zhukovskii, N. E. "Mekhanika v Moskovskom universiteta za poslednee piatidesiatiletie." in his *Polnoe sobranie sochinenii*. Moscow: Glavnaia Redaktsiia Aviatsionnoi Literatury, 1937. Vol. 9, pp. 203–11.

Znakomye. Al'bom M. I. Semevskogo. St. Petersburg: V. S. Balasheva, 1888.

INDEX

abelian and elliptic integrals and functions, 122, 144, 148n, 234, 242–243, 247, 253–255
Academy of Sciences, Berlin, 188
  Bologna, 223n
  French, 6, 186, 197–198, 206–209, 211, 240, 245, 269–270
  Russian, xii, 6, 10, 118–119, 196, 221–223, 226–228, 231–232, 235n, 270
  Swedish, 6, 161, 193–194, 235n
Acta Mathematica, xvii–xviii, 6, 161, 196–197, 203, 208, 228, 247, 270, 272
Adelung, Sophie, 78, 152n, 277
Agnesi, Maria Gaetana, 223n
Alexander II, Tsar, 152
Alexander III, Tsar, 202
Andrews, George, 272
Aniuta (see Korvin-Krukovskaia, Anna)
Archibald, R. C., 166
Armfeldt, Natalia, 92, 96–97, 258
At the Exhibition (Kovalevskaia), 263–264
Babukhin, A. I., 148
Baer, Karl Ernst von, xii
Bakunin, Mikhail, 71
Bassi, Laura, 223n
Bell, Eric Temple, 129, 278–279
Beltrami, Eugenio, 217–218
Berlin Academy of Sciences, 188
Berlin University, 100–101, 188–189, 245, 271
Bernoulli, Daniel, xii
Bernoulli, Nicolaus, xii
Bertrand, Joseph, 198, 206, 209, 217, 249, 252
Biermann, Kurt, 117
Bismarck, Otto von, 104
Bjerknes, Carl, 217–218
Blackwell, Elizabeth, 65
Blanqui, Louis, 104
Boborykin, P. D., 213
Bogdanova, Maria, 65
Bokov, Petr, 62, 72, 76n, 77, 81, 146

Bokova-Sechenova, Maria, 4, 24, 62, 65, 72, 76–77, 169, 171, 202–203, 236n, 270
Bologna Academy of Sciences, 223n
Borisiak, A. A., 72, 128n
Brandes, Georg, 135
Brandt, E. K., 63
Branting, Karl Hjalmar, 218
Brehm, Jacob, 191
Briullova, Aunt Sofia, 60, 78
Brower, Daniel, 61n, 65n
Büchner, Ludwig, 57
Buckle, Henry Thomas, 57
Bugaev, N. V., 148
Buinitskii, 51–53
Bulwer-Lytton, E., 33
Buniakovskii, V. Ia., 222n, 226, 228
Bunsen, R. Wilhelm, 7, 89, 245
Butlerov, A. M., xiii, 4n, 127, 145
Buys, Mutiara, 243n, 255n
Cauchy, Augustin, 122, 240–241
Cauchy-Kovalevskaia Theorem, 122, 240–242, 245, 253
censorship in Russia, 51n, 75, 230, 257–258, 260–261, 267
Chebyshev, P. L., 126–127, 131, 133, 185, 197, 222–223, 226–228, 246–248
Chekhov, Anton, 7, 213, 230, 235
Chernyshevskaia, Olga, 83, 94n, 139, 263
Chernyshevskii, Nikolai, 7, 61–62, 64, 70, 82–83, 111, 139, 220, 229, 233–234, 259, 263
Cooke, Roger, 143n, 203n, 279
Crimean War, ix, xii–xiii, 11n, 14, 61
Curie, Maria Sklodowska, 53, 186, 271
Darboux, Gaston, 162, 198, 240–241
Darwin, Charles, ix, 7, 57, 90, 115, 139
Davidov, A. Iu., 148
Davitashvili, L. Sh., 71–72, 128n
DeRoberti, E. V., 214
Dirichlet, Paul, 232, 253
Dollo, Louis, 172n
Dostoevskaia, Anna (Snitkina), 45n, 202

299

Index

Index

Panteleev, L. F., 62, 68, 125n, 134, 136
Paris Commune of 1871, 31n, 97, 104–
    106, 108–111, 125, 200, 233, 234n
Paris Mathematical Society, 162
Pavlovskii, I. Ia., 262
Perrot, Joseph, 165–166
Perrot, Zinaida (Zina), 165–167, 172
Petrushevskii, F. F., 63, 83
Picard, Emile, xvii, 162, 185, 198, 213,
    232, 250, 252, 254–255
Pirogov, N. I., 9, 11
Pisarev, D. I., 61–62
Pobedonostsev, K. P., 202
Poincaré, Henri, xvii, 7, 162, 198, 202,
    213, 219, 241, 244, 249–250, 252–253
Poisson, Siméon, 242
Poland,
    "Proletariat" Party, 201, 202n
    revolutionary movement in, 163–165,
        181, 184, 201–202, 223–224, 229–230
    uprising of 1863, 19–20, 27–28, 50–53,
        69
Polubarinova-Kochina, P. Ia., 60, 129,
    135n, 244, 279
populism (see also nihilism, sixties), 62,
    125n, 261–263
Prix Bordin, 6, 177, 201, 203, 206, 208–
    209, 211–212, 217, 240, 270, 274
pseudoscience, 57, 91, 258, 265
Pypin, Aleksander, 83, 139, 229, 257
Ragozin, Leonid, 146, 152, 160–161, 168–
    173
Ragozin, Viktor, 146–147, 160–161, 168–
    173
Ragozin oil company, 146–147, 150–152,
    160–161, 168–173
Rappaport, Karen D., 278–279
refraction of light in a crystalline
    medium, 143, 153, 155, 168, 172–174,
    190, 244
revolution (rotation) of a solid (rigid)
    body about a fixed point (the
    Kovalevskaia Top), 153, 197–198, 206–
    209, 242–243, 253
revolutionary émigré community, 163–
    164, 214–215, 223–224, 258
Reznik, Semen, 129n, 146n
Riemann, Georg, 249
Rodde, Dorothea von Schlözer, 122n
Rozhdestvenskii, I. G., 68–69, 72
Runge, Carl, 185, 190, 252
Russia,
    Academy of Sciences of, xii, 6, 10, 118–
        119, 196, 221–223, 226–228, 231–232,
        235n, 270
    Congresses of Natural Scientists in, 144,
        174, 243n, 246

intelligentsia of, ix–xvi, 9, 14, 35–36, 56–
    58, 60–65, 75–76, 81–83, 126–128, 133–
    135, 183, 208, 214–215, 235–237, 257–
    258, 261, 263
    revolution of 1905, 261
    revolution of 1917, 219n, 261
    scientific community in, xi–xviii, 118–
        119, 126–127, 144, 171, 174
    women in, ix–xvi, 39, 55–57, 61–65, 73,
        91–92, 100, 134n, 140, 236–237, 263,
        270–271
Russian émigrés abroad, 183, 214–215,
    218–219, 223–224
Russkoe slovo (The Russian Word), 38
Saburov, A. A., 148
Saltykov-Shchedrin, M. E., 134–135,
    214–215
Saturn's rings, on the shape of, 122, 243–
    244
Schwarz, Hermann, 114, 117, 121, 252
Science and scientists (see also individual
    listings and women in science), x–
    xviii, 2–8, 20–23, 36, 39, 46–50, 53–54,
    57–58, 69, 73–75, 88–90, 100–101, 115,
    118–119, 126–128, 144–145, 150, 168n,
    171–172, 180, 194, 196, 208, 212–213,
    221–223, 231, 235, 245, 255, 269–275
Sechenov, Ivan, xiii, 36, 72, 76, 81, 118,
    145–146, 202–203, 213, 236n
Selivanov, D. F., 185
Semevskii, M. I., 15n, 21n, 23, 35–37, 53
Senkovskii, O. A., 9
Severnyi vestnik (Northern Messenger),
    174, 198, 230
Shabelskaia (Montvid), A. S., 231
Shelgunov, N. V., 69–70, 82
Shelgunova, Liudmila (Mikhaelis), 70,
    73, 94n
Shishkin, N., 191n
Shtraikh, S. Ia., 71, 133, 279
Shubert, F. F. (1789–1865), 10
Shubert, Fedor F. (1831–1877), 21–22,
    132
Shubert, F. I., 10
Siberia, exile system in, 41, 52, 62, 223–
    224, 261–262
sixties (see also nihilism, populism),
    men of the, ix–xvi, 81–82
    people of the, ix–xvi, 82, 133, 263, 270–
        271, 274–275
    women of the, ix–xvi, 2, 4–5, 57–58, 62,
        65n, 100, 263, 270, 274–275
Skabichevskii, A. M., 63n
Sleptsov, V. A., 39n
Smith, Margarita, 16, 21n, 24–25, 31–32,
    40, 46, 51–52
Snegirev, V. F., 148

304